Vicia faba:

PHYSIOLOGY AND BREEDING

WORLD CROPS:
PRODUCTION, UTILIZATION, AND DESCRIPTION

volume 4

Other volumes in this series:

1. Stanton WR, Flach, M, eds: SAGO The equatorial swamp as a natural resource. 1980. ISBN 90-247-2470-8.
2. Pollmer WG, Phipps, RH, eds: Improvement of quality traits of maize for grain and silage use. 1980. ISBN 90-247-2289-6.
3. Bond DA, ed: Vicia faba: Feeding value, processing and viruses. 1980. ISBN 90-247-2362-0.

Series ISBN: 90-247-2263-2

Vicia faba:

PHYSIOLOGY AND BREEDING

Proceedings of a Seminar in the EEC Programme of Coordination of Research on the Improvement of the Production of Plant Proteins, organised by the Centrum voor Agrobiologisch Onderzoek (Centre for Agrobiological Research), Wageningen, The Netherlands, held in Wageningen, June 24-26, 1980

Sponsored by the Commission of the European Communities, Directorate-General for Agriculture, Coordination of Agricultural Research

edited by

R. THOMPSON

1981

MARTINUS NIJHOFF PUBLISHERS
THE HAGUE / BOSTON / LONDON
for the
COMMISSION OF THE EUROPEAN COMMUNITIES

Distributors:

for the United States and Canada

Kluwer Boston, Inc.
190 Old Derby Street
Hingham, MA 02043
USA

for all other countries

Kluwer Academic Publishers Group
Distribution Center
P.O. Box 322
3300 AH Dordrecht
The Netherlands

This volume is listed in the Library of Congress Cataloging in Publication Data

ISBN 90-247-2496-1 (this volume)
ISBN 90-247-2263-2 (series)

Publication arranged by
Commission of the European Communities,
Directorate-General Information Market and Innovation

EUR 6895 EN

Proceedings prepared by:
Janssen Services, 14 The Quay, Lower Thames Street, London, EC3R 6BU, UK 

PRINTED IN THE NETHERLANDS

CONTENTS

PREFACE

This is the third major publication on *Vicia faba* reporting proceedings of seminars organised through the Commission of the European Communities in the context of the EEC Common Research Programme on Plant Protein Improvement.

The previous two volumes report proceedings from the seminars in Bari in 1978 (Some current research on *Vicia faba* in Western Europe) and in Cambridge in 1979 (*Vicia faba:* Feeding value, processing and viruses).

The theme of this seminar, held in Wageningen on 24th - 26th June, was selected to examine various aspects of plant physiology, especially in relation to their potential contribution to plant breeding. Areas identified for discussion included those to elucidate particular responses of the plant to the environment and, in the broader concept, an examination of the combined responses of ideotypes required to exploit fully the range of environments in which *Vicia faba* is grown.

Increased pea breeding in Europe in recent years justified an assessment of the projected progress of the crop for dry seed production and two papers were invited on this topic.

Participants visited Cebeco at Lelystad where, after hearing of the involvement in field beans and peas, the advanced selections in field plots were examined and discussed. On the same day visits were also made to the SVP experimental farm and to the CABO experimental farm at de Eest, to discuss the plant breeding/physiology experiments on field beans and peas.

The seminar was organised by Dr. G. Dantuma and Ir. R.J. Heringa together with colleagues to whom much credit is due for its great success.

OPENING REMARKS

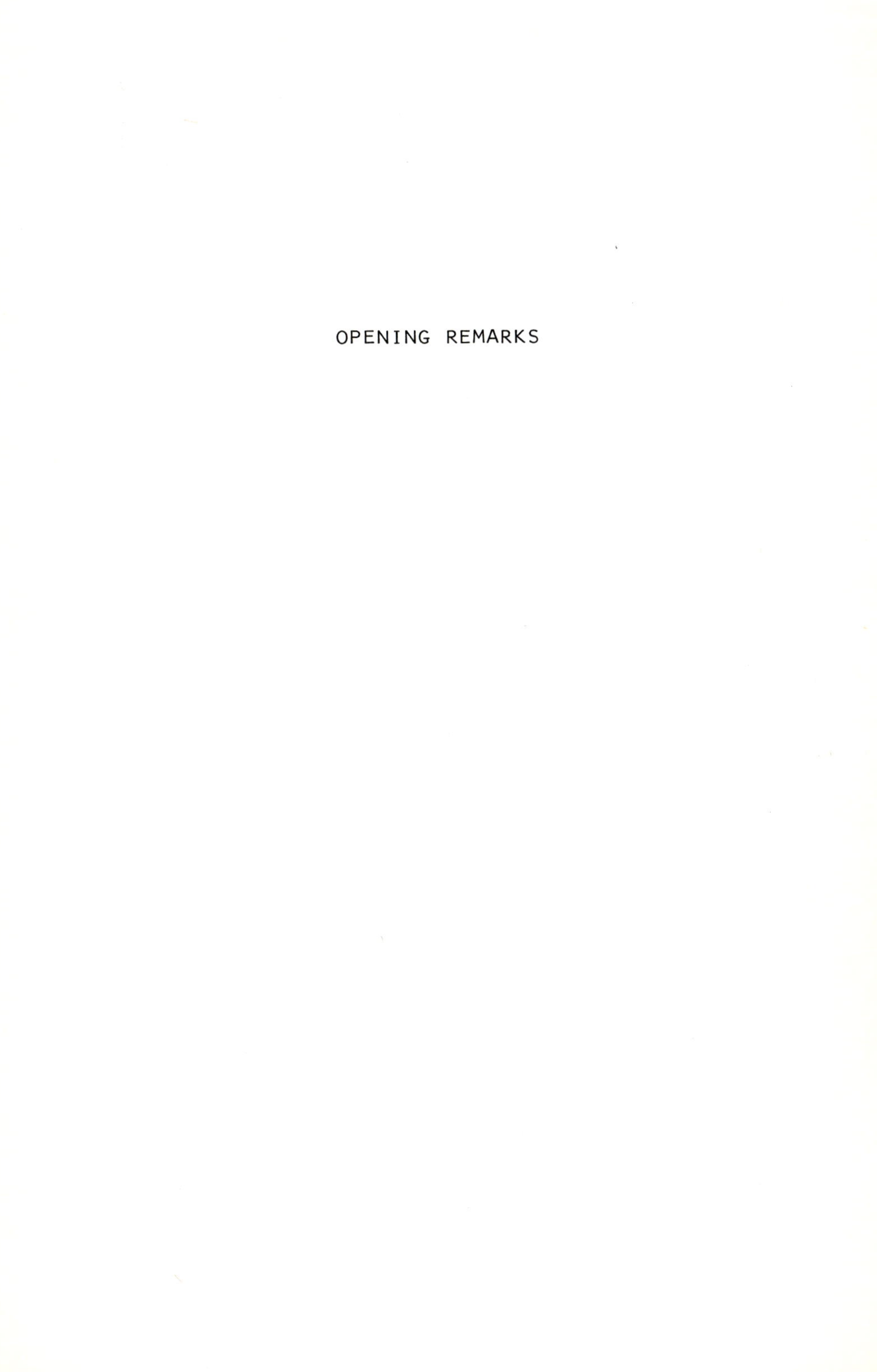

Ir. R. Mulder
Ministry of Agriculture,
Division for Agricultural Research,
Wageningen, The Netherlands.

I would like to say a few words at the opening of this Seminar on behalf of my colleaques who have organised it. I have the pleasure to welcome you all most heartily to this IAC building where we are able to hold the seminar on seed legumes. I would like especially to extend a welcome to Dr. Prendergast from the Commission. As you all know, the Commission has supported the research programme on plant proteins for several years now. It has done so mainly through two types of support. First, it supports national research projects of common interest, and secondly, it provides support by co-ordination activities, through workshops, seminars, and so on. The financial support for national projects of common interest has undoubtedly stimulated national research efforts in this field of research which is aimed at a high degree of independence in the protein supply of livestock in Europe. In my opinion these common programmes, or contracts, would not have been nearly so effective without the sustaining co-ordination activities. For example, thanks to these co-ordinating activities, this group on seed legumes has already been able to meet four times. During these meetings members of the group have rapidly come to know each other, and they have succeeded in collaborating efficiently and pleasantly by exchanging views, ideas, information, knowledge and materials. As a result, good progress has been made, for example, in analysing the main yield limiting factors of European crops and in identifying desirable genetic traits for plant breeding.

However, for maximum effectiveness of EEC support, a certain equilibrium should be maintained between contracts and co-ordination activities. No contracts means less effort on the national level, fewer results and little to discuss in workshops and seminars. No co-ordination means no common use of results. At times when funds are short, as at present, there may be a tendency towards fewer contracts and more co-ordination. I do not think this would be the right development. Therefore, I sincerely hope that the Commission, together with the Expert Group, will be able to maintain the best balance between contracts and co-ordination, in spite of budgetary difficulties. I also hope that the results from the various research groups, including this one on seed legumes, will clearly indicate in which direction a possible new research programme on plant protein should be directed. We know that the Expert Group has already started to discuss such a new programme. I also hope that the discussions in this seminar will contribute to this.

Finally, may I wish you all a very successful seminar at Wageningen, good weather during your field trips, and last, but not least, a pleasant stay in the Netherlands.

Thank you.

SESSION I

PHYSIOLOGY

Chairman: D.A. Lawes

PHYSIOLOGICAL ASPECTS OF GROWTH AND DEVELOPMENT OF *Vicia faba*

H.M. Dekhuijzen[1], D.R. Verkerke[1] and A. Houwers[2]

[1]Centre for Agrobiological Research, Wageningen, The Netherlands.

[2]Department of Microbiology, Agricultural University, Wageningen, The Netherlands.

ABSTRACT

This study was initiated to establish the pattern of nitrate uptake, nitrate reduction and nitrogen fixation by Vicia faba *at an experimental farm. Nitrate uptake was estimated by measuring the nitrate reductase activity (NRA) on the youngest developing leaf. Nitrogen fixation (NF) was measured with the acetylene reduction assay on root systems. The highest NRA occurred at the end of flowering two weeks before the highest NF was recorded during mid pod filling. NRA was correlated with the rainfall. After a period of rainfall the NRA decreased, whereas the NF increased. From these results it was inferred that the decrease in NF activity during the vegetative period up to the end of flowering was due to flooding or rainfall resulted in a higher uptake of nitrate which caused a reduction of NF activity. Nitrate fertilisation did not increase seed yield and suppressed NF.*

Growth and development of plants grown hydroponically in a growth chamber depended strongly on temperature. Developmental stages were plotted against data on accumulated daily temperatures in a growth chamber experiment. An attempt was made to use these kind of data to predict certain developmental growth stages under field conditions.

INTRODUCTION

One of the characteristics of spring sown *Vicia faba* is its large and unpredictable variation in yield in different years. Some factors which are known to affect yield are soil moisture, temperature, light and disease. These factors affect fixation of carbon and nitrogen. Attempts to increase yield by applying fertiliser nitrogen have generally resulted in only small non-economic effects on the yield (Day et al., 1979; Richards and Soper, 1979). No satisfactory physiological explanation for these results has been reported. Seasonal profiles of nitrate reductase activity in soyabean leaves, which indicates nitrogen uptake levels have been reported by Harper (1974) and Thibodeau and Jaworski (1975). Shortly after full bloom nitrate reduction declined rapidly. Nodule growth and nitrogen fixation in soil grown plants occurred largely after flowering, whereas the major nitrogen fixation was observed during the period of early pod fill.

The present study was initiated to establish the pattern of nitrate uptake, nitrate reduction and nitrogen fixation by *Vicia faba* under field conditions. In addition, the effects of day and night temperature on growth and nitrate uptake was determined in hydroponically grown plants in a growth chamber.

MATERIAL AND METHODS

Field grown plants

Seeds of *Vicia faba*, variety Minica (20 plants/m^2) were sown at an experimental farm, De Bouwing at Randwijk near Wageningen on heavy (60%) clay (Dantuma and Klein Hulze, 1979). Soil was fertilised with P and K each at 240 kg/ha. Seeds were sown in rows 50 cm apart on April 17, 1978. The field was subdivided in plots of 2 x 7.5 m^2. Some plots were fertilised with 200 kg N/ha on 16 May 1978 at the three leaf stage of the plants, whereas other plots were fertilised with 200 kg N/ha on 27 June 1978, at the end of flowering. Plots which were not fertilised with N served as controls. Three plots were sampled each week (3 x 20 plants) during the growing season.

Plants grown in hydroponic culture

Plants were grown in plastic pots of 10 l filled with 8.5 l of a nutrient solution in a growth chamber. The nutrient solution was renewed 3 times each week because otherwise the pH increased too rapidly. In this way pH was kept at 6.5 - 7.0. During replacement of the solution, root and shoot weight were determined. Root weight was estimated by subtracting weight of the total plant minus weight of the plant with the roots immersed in water. The uptake of water from the pots was measured and aliquots were taken for determination of nitrate uptake. The nutrient solution contained KH_2PO_4 135 mg, $MgSO_4.7H_2O$ 498 mg, K_2SO_4 252 mg, Ca $(NO_3)_2 4H_2O$ 1 074 mg and KNO_3 282 mg/l deionised water. This solution was supplemented with iron chelate, 2.5 mg/l and microelements. One group of 4 plants was grown at 16°C continuously (14 h light) and another at 16°C day and night up to leaf stage 6 and then subjected to a regime of 26°C for 14 h (light period) and 16°C during the dark period of 10 h. Plants received an irradiance of 80 Wm^2 (300 - 800 nm). Light sources consisted of Philips HPL/N 400 W and incandescent lamps.

Nitrate uptake

Nitrate uptake was measured 3 times a week with a nitrate electrode.

Nitrate reductase activity (NRA)

The intact tissue assay of Jones and Sheard (1977) with minor modifications was used. Leaf discs (2 - 3) with a diameter of 1 cm were cut with a cork borer from the youngest developing top leaf at 9 o'clock in the morning. Leaf discs (0.5 g per plot of 20 plants) were infiltrated under vacuo (N_2) in 10 ml of a medium containing 50 mM KNO_3, 1% n-propanol, 100 mM P-buffer and 40 µg/ml chloromphenicol at 27°C. Nitrate released into the medium after 30 and 60 minutes was determined with sulfanil amide and 0.02% N-1-naphthylethylene diamine. Optical density was measured at 540 nm.

Nitrogen determination

Kjeldahl nitrogen analyses were performed on samples of field grown plants (6 plants of each plot of 20 plants).

Nitrogen fixation

The acetylene reduction method as described by Thibodeau and Jaworski (1975) was used.

RESULTS

Field experiment

Plant development is shown in Figure 1, whereas Figure 2 gives data on the nitrogen increase in the above ground parts during the growing season. Nitrate fertilisation on 16 May 1978 slightly increased the total amount of N in the plants.

Diurnal variation of nitrate reductase activity (NRA) was measured every hour on leaf number 2 and 4 (youngest) under field conditions on plants two weeks before flowering. NRA increased in the youngest leaf from 2 μmol $NO_2 \cdot g^{-1} FW \cdot h^{-1}$ prior to 9.00 o'clock in the morning to a level of 3 μmol NO_2 between noon and 14.00 (Figure 3). Similar values have been found for soyabeans by Thibodeau and Jaworski (1975). For convenience field sampling was carried out during the whole season once a week at 9.00. Nitrogen fixation was measured half an hour later.

NRA fluctuated during the season. Application of nitrate on 16 May 1978 increased NRA slightly (Figure 4) and the total amount of N in the plants. The second application of nitrate at 27 June 1978 increased NRA but not the amount of N in the above ground parts. Nitrogen fixation (NF) also fluctuated strongly during the growing season. The highest activity

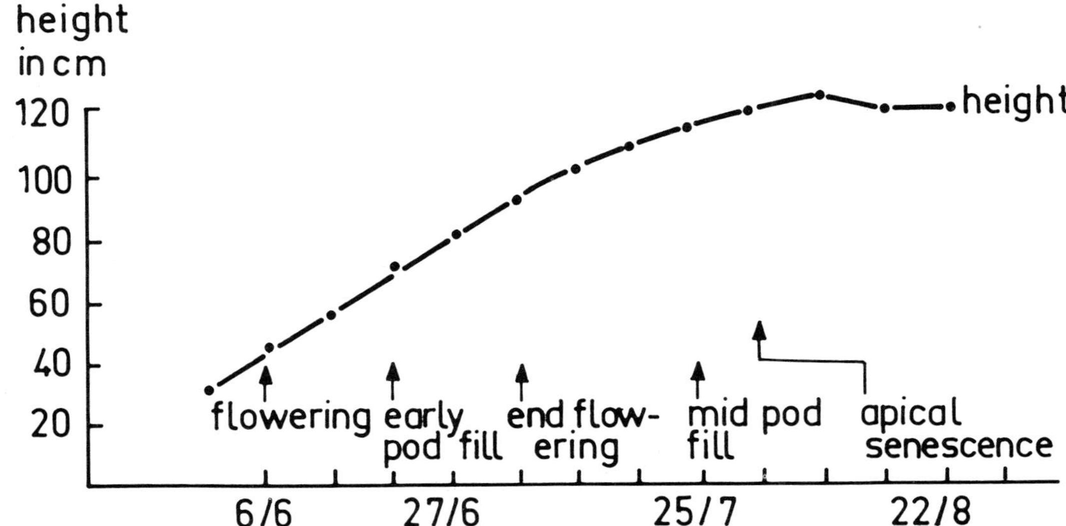

Fig. 1. Growth (total length) of field beans during the growing season of 1978.

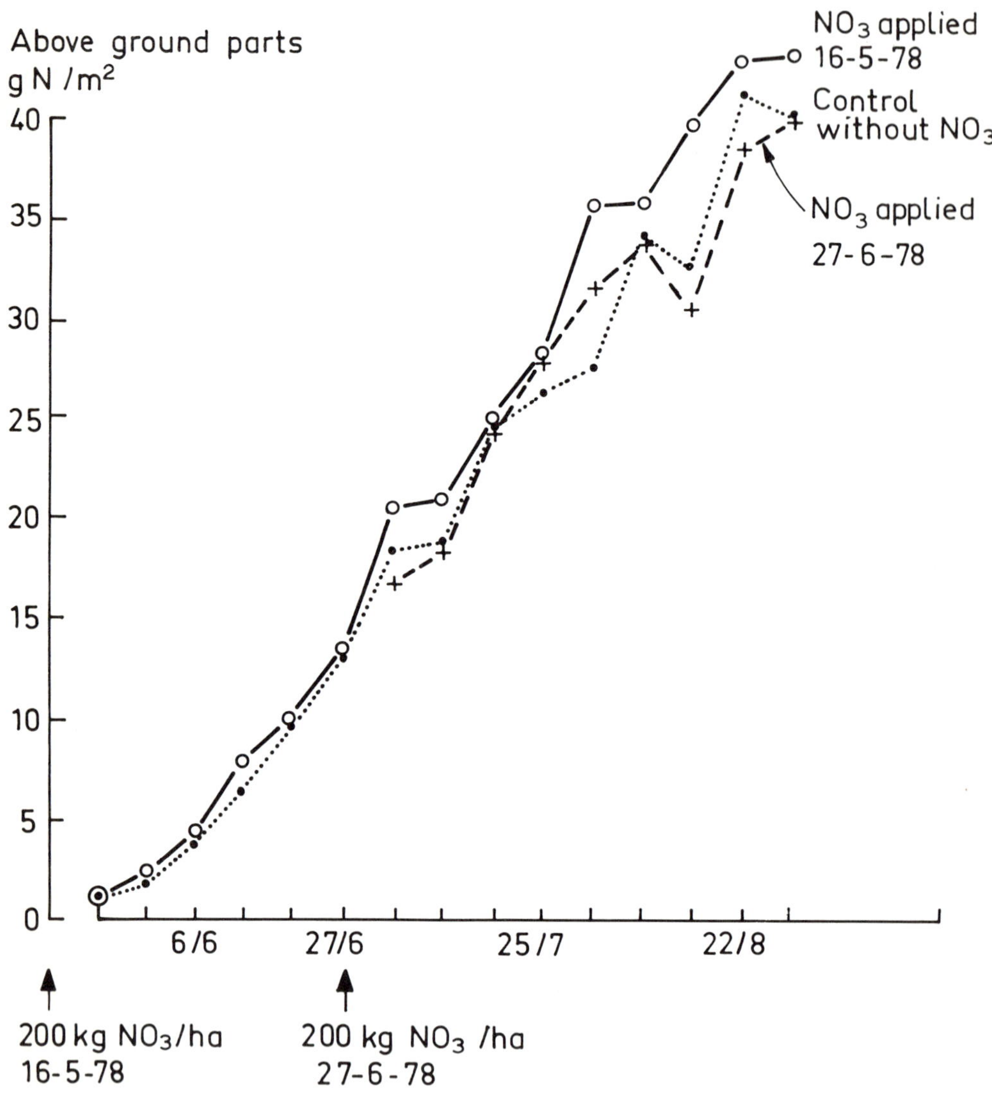

Fig. 2. Effect of nitrate fertiliser on total N in *Vicia faba*. Nitrate was applied on 16 May 1978 and on another plot on 27 June 1978.

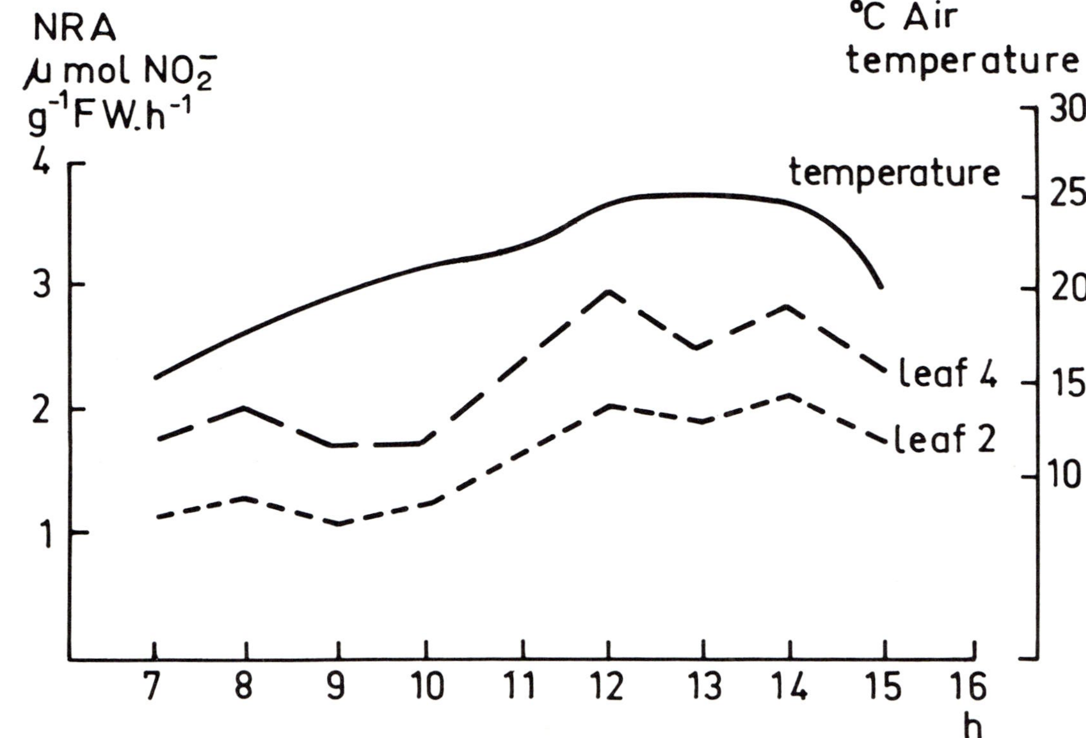

Fig. 3. Nitrate reductase activity (NRA) of leaf 2 and 4 between 7.00 and 15.00 h

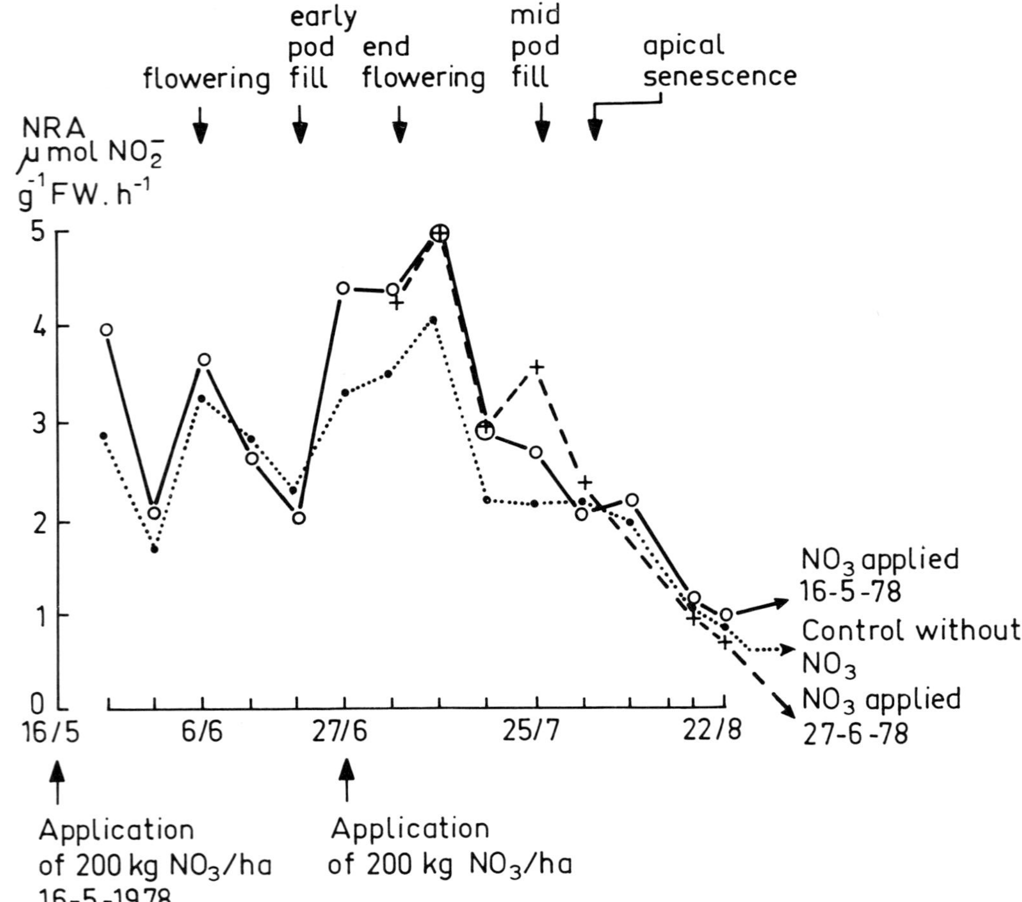

Fig. 4. Effect of application of nitrate at 16 May 1978 and at 27 June 1978 on nitrate reductase activity (NRA).

occurred on 25 July 1978 during the mid pod filling period which was 2 weeks later than the highest recorded NRA at the end of flowering (Figure 5). The latter figure also shows the well known fact that nitrate fertilisation reduces nitrogen fixation. An attempt was made to estimate the contribution by NF and NRA during flowering. Although the total activity of NRA+NF over-estimated the actual N uptake, it is clear from Table 1 that the contribution by NF was much smaller than by NRA on this soil type. Nitrate fertilisation inhibited NF by about 50%. The average estimated N uptake of the plants was 6.0 kg $N.ha^{-1}$ against a mean actual uptake of 4.5 kg $N.ha^{-1}.day^{-1}$.

TABLE 1

CONTRIBUTION OF NITRATE UPTAKE AND NITROGEN FIXATION TO TOTAL N UPTAKE DURING FLOWERING FROM 6 JUNE 1978 TO 13 JUNE 1978

NO_3 fertilisation	Contributed	Uptake of N from 6 June 1978 to 13 June 1978	
		Estimated g N/m^2	Actual N uptake g N/m^2
None	NF	82	
	NRA	331	
	Total	413	322
200 kg^{-1} ha^{-1}	NF	47	
	NRA	386	
	Total	433	322

NRA was found to be correlated with the rainfall during the preceding week. After a period of rainfall the NRA decreased whereas the NF increased (Figure 6). From the results it may be inferred that the decrease in NF activity during the vegetative period up to the end of flowering is either due to flooding or rainfall and resulted in a higher uptake of nitrate which reduced NF activity. The nitrate is either

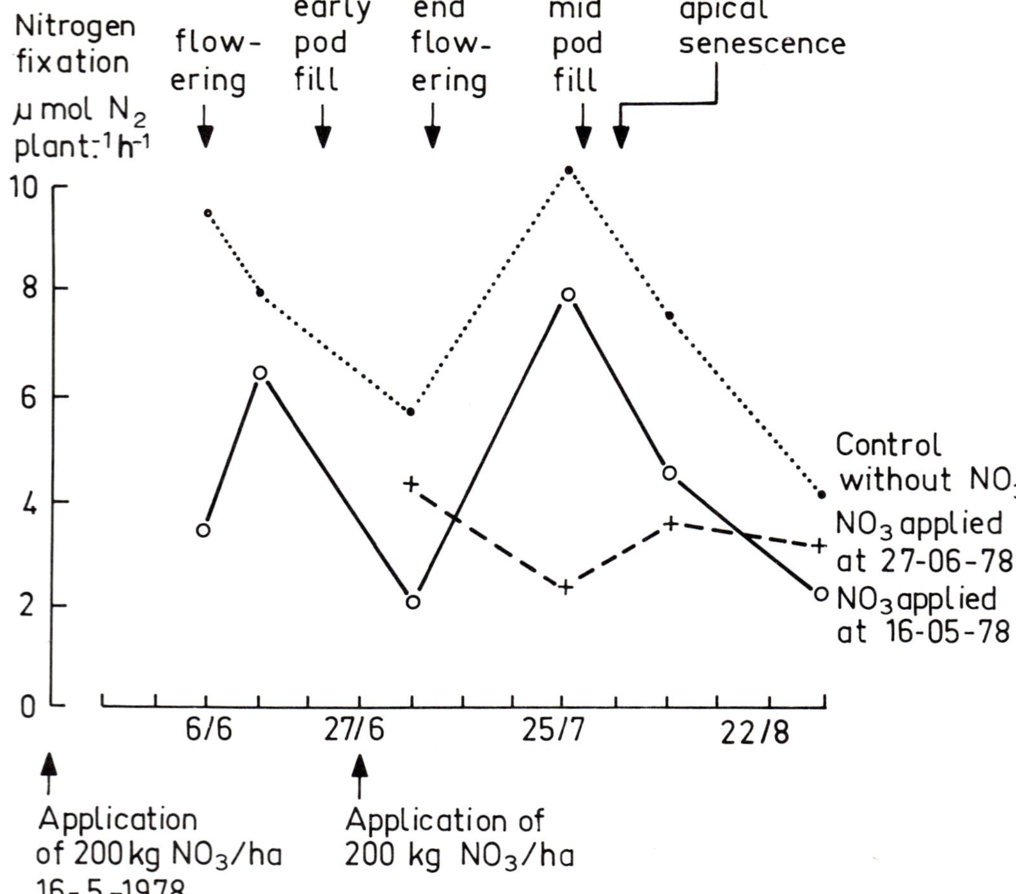

Fig. 5. Effect of nitrate fertiliser on nitrogen fixation (acetylene reduction assay). Nitrate was applied to two different plots, one at 16 May 1978 and one at 27 June 1978.

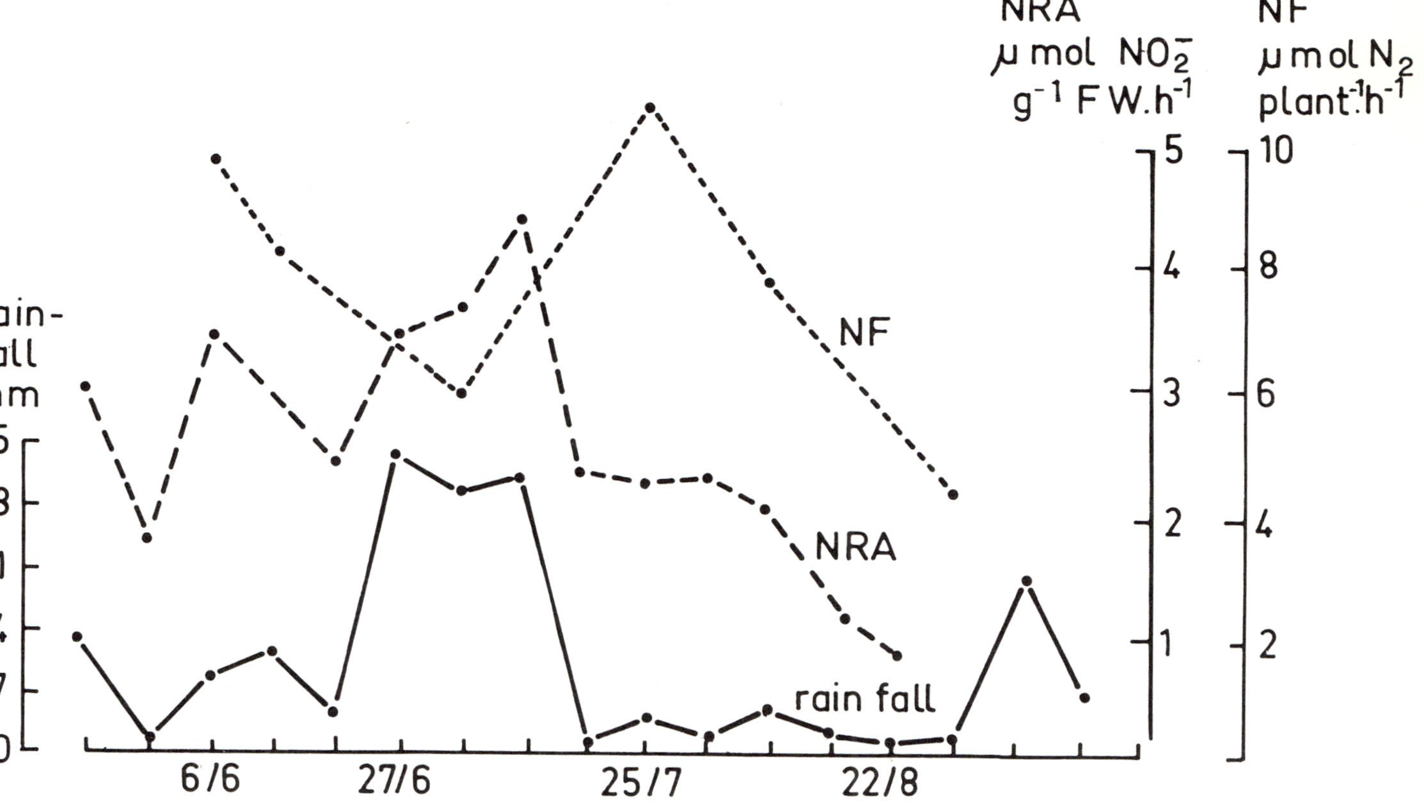

Fig. 6. Nitrate reductase activity (NRA) and nitrogen fixation (NF) in relation to rainfall during the growing season of 1978 (without NO_3).

free available in the soil or is set free from organic material by micro-organisms during the wet period.

Effect of day and night temperature on growth and nitrate uptake

Total yield differs from one variety to another, but also the harvest indices may vary from one locality to another. Dantuma and Klein Hulze (1979) in the Netherlands found harvest indices of 65% for variety Minica, whereas Thompson and Taylor (1979) in Scotland reported 31% for the variety Herz Freya grown in the field where constraints such as water supply and diseases were minimised. Besides differences in varieties, climatic differences such as temperature, irradiance and photo-period may play an important role in determining the final yield. We choose one factor, i.e. day and night temperature as a possible factor affecting growth and yield. Sprent et al. (1977) mentioned the factor temperature but did not study its effect in detail. Abdalla and Fischbeck (1978) and Said et al. (1967) reported a strong effect of day and night temperature on seed yield. Fresh weight of above ground parts as well as of the roots of plants grown continously at 16°C (Figure 7) increased linearly between 50 and 80 days from planting. A linear relationship exists between fresh weight of the above ground parts and cumulative nitrate uptake from the nutrient solution (Figure 8). At this temperature neither the leaves nor the pods senesced during the experimental period of 91 days.

Growth and development of plants which were subjected to a night-day rhythm of 16 - 26°C, 49 days after planting, differed completely from those grown at 16°C both day and night. Root growth decreased about day 55 and stopped after day 63, whereas NO_3 uptake proceeded at least up to day 85. After day 85 fresh weight of the above ground parts decreased.

Pods started to swell rapidly when plants were transferred from 16°C to a regime of 16°/26°C, 49 days after planting

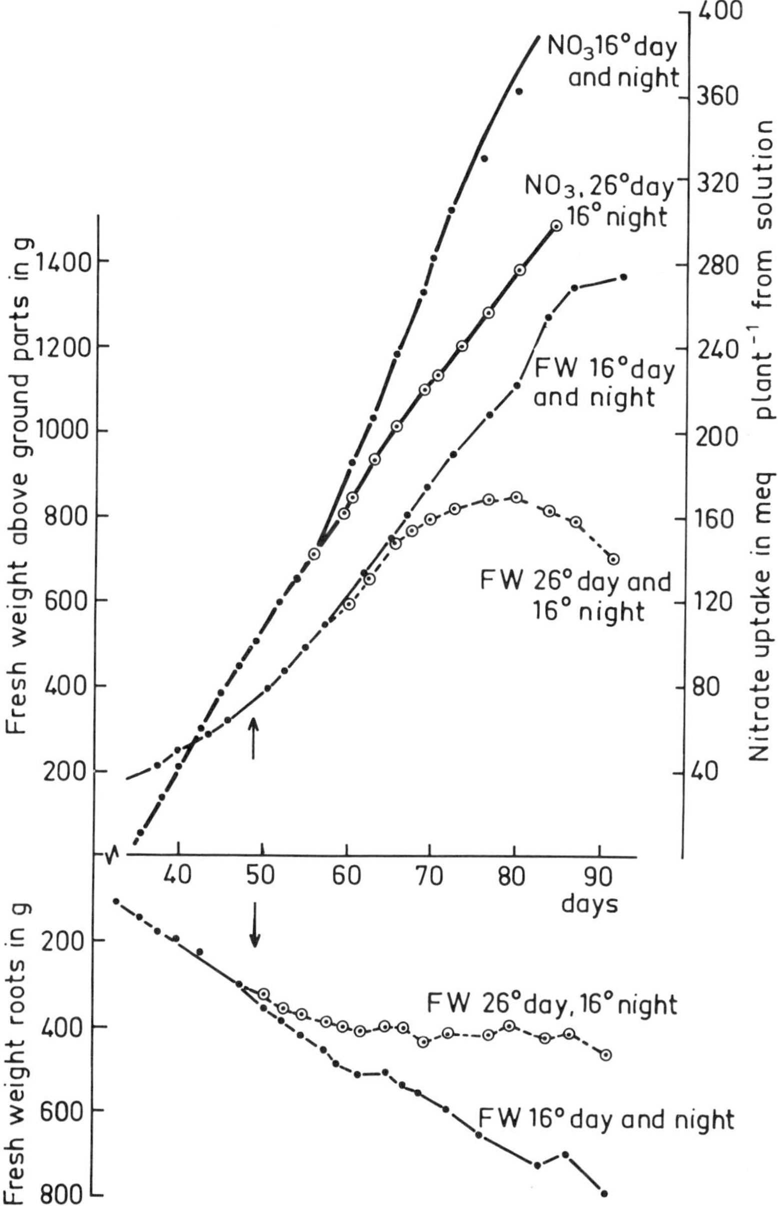

Fig. 7. Effect of day and night temperature on growth and nitrate uptake of *Vicia faba*. Up to 49 days (arrow) after planting all plants were hydroponically grown at 16°C during the day and night. Thereafter plants were split into two groups, one growing continuously at 16°C and another at 26°C during the day and at 16°C during the night period of 10 hours.

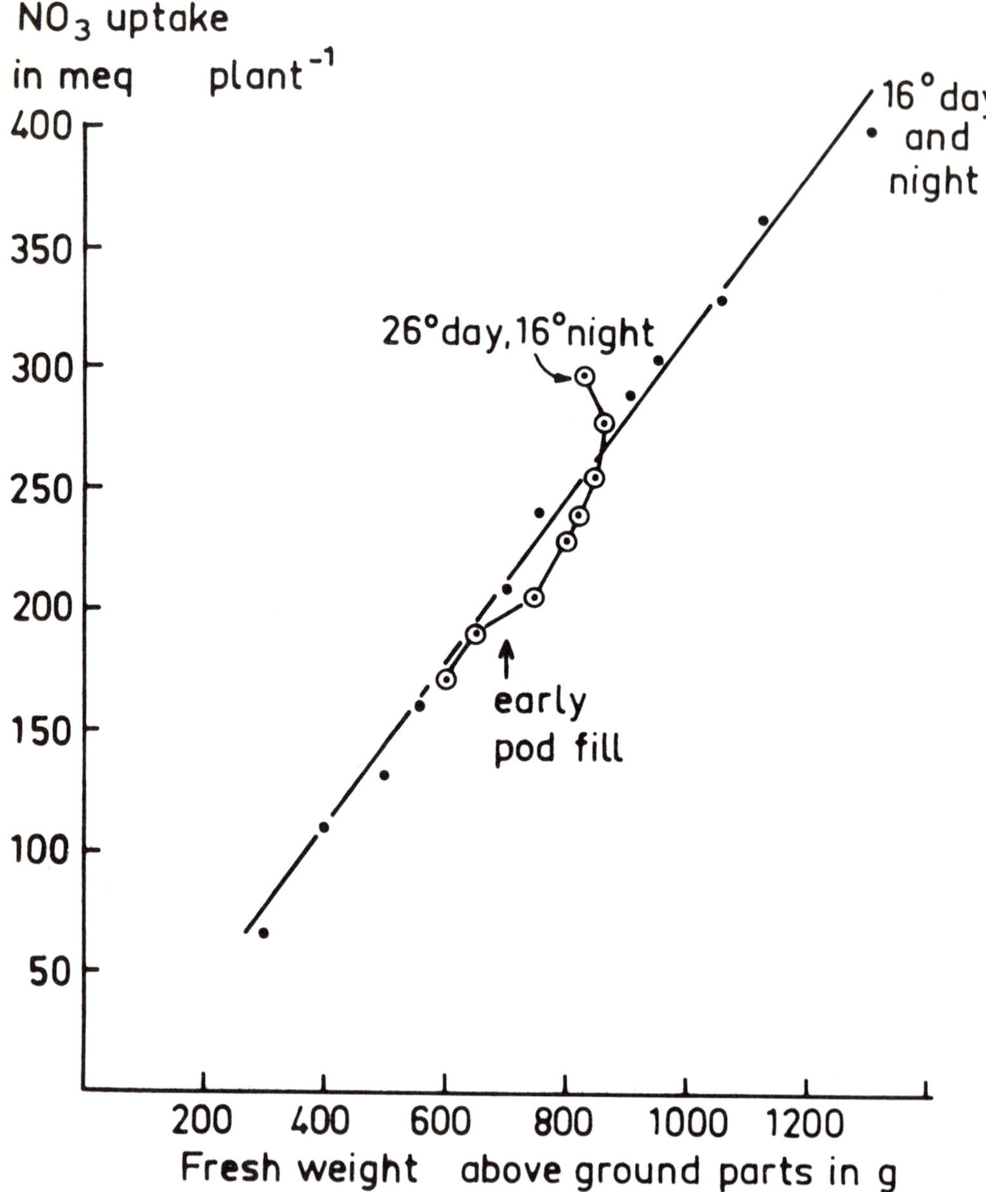

Fig. 8. Relationship between fresh weight and nitrate uptake of plants grown at 16°C (⊙) or at 26°C during the day and 16°C during the night (●).

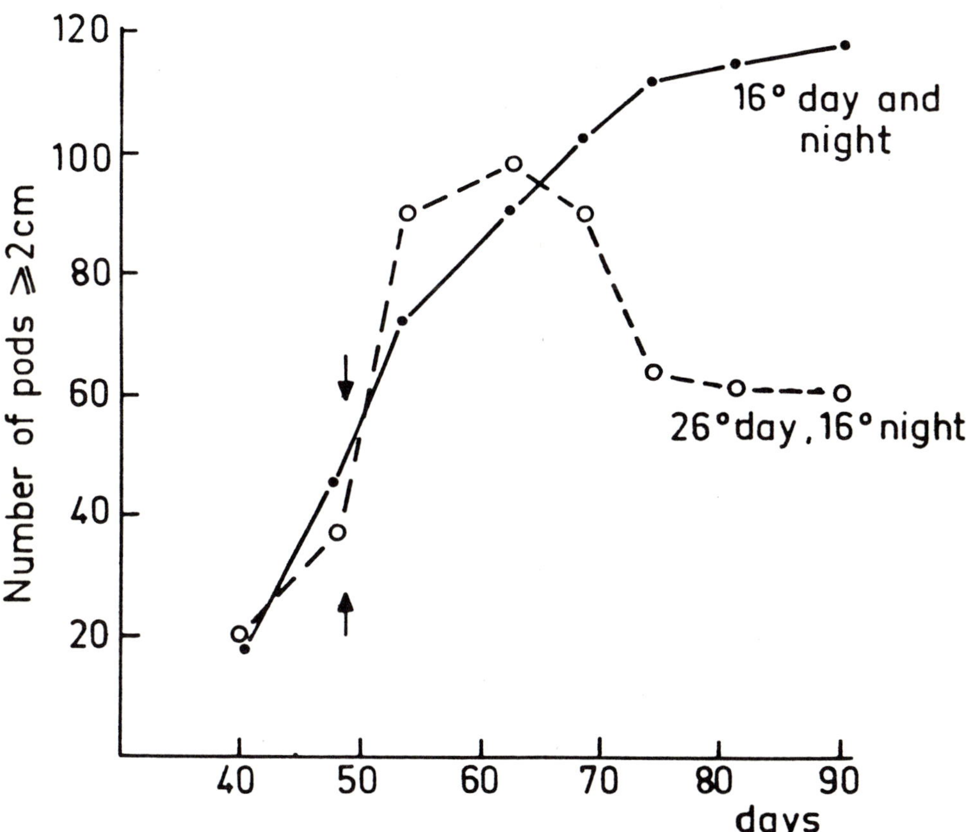

Fig. 9. Effect of day and night temperature on the number of pods larger than 2 cm. Conditions as in Figure 7.

(Figure 9). However, newly formed flowers and pods aborted at a regime of $16^{o}/26^{o}C$ much more rapidly than at $16^{o}C$.

Whereas, senescence of the oldest leaves of plants grown at $16^{o}C$ did not occur, those growing at $16^{o}/26^{o}C$ started to senesce between day 63 and 66 after planting. Pod walls of the oldest pods of plants grown at $16^{o}/26^{o}C$ started to senesce 84 days after planting. Although root growth terminated after 63 days, NO_3 uptake proceeded up to day 85. Figure 8 clearly shows that plants grown at $16^{o}/26^{o}C$ took up NO_3 even after fresh weight (850 g) of the above ground parts did not increase further (from 80 days after planting). Sprent (1977) and Gehriger and Keller (1979) reported that under field conditions root growth stopped when the pods started to swell. According to Gehriger and Keller (1979) root weight remained constant from the start of pod swelling until harvest.

DISCUSSION

Results with *Vicia faba* grown under field conditions showed that the optimal NRA preceded that of NF activity. Thibodeau and Jaworski (1975) obtained the same results with soyabeans but did not mention weekly rainfall. Only data on monthly rainfall are given. Our results clearly showed a correlation between NRA and weekly recorded rainfall. It seems therefore most likely that abundant water stimulated mineralisation and nitrate uptake from the soil which resulted in a higher NRA and simultaneously a lower NF activity. These results are consistent with those of Richards and Soper (1979) showing that beans grown on soil rich in organic material preferentially take up available nitrate rather than symbiotically fix nitrogen. Addition of N fertiliser resulted in a reduction of symbiotic fixation with no corresponding economical increase of the seed yield.

It seems remarkable that data on seed yield and on redistribution of N in *Vicia faba* are so variable. Richards and Soper (1979) in Canada reported a seed yield of 2 000 kg/ha and a N-index of 60%, whereas Dantuma and Klein Hulze (1979) in the Netherlands near Wageningen obtained 5 780 kg/ha and a N-index of 90% with the variety Minica. Sprent et al. (1977) and Sprent and Bradford (1977) found in Scotland a seed yield of 6 712 kg/ha in 1974 and a N-index of 59% for the variety Maris Bead. Cooper et al. (1976) grew Maris Bead in sand with a low concentration of nitrate (3.25 meq/l) in a cool greenhouse (Table 2); plants nodulated and were pulse labelled with $^{15}NO_3$ through the growing season from May to November. $^{15}NO_3$ was taken up and incorporated into the seeds even when fed to plants with green pods between 3 - 9 September. Quantitatively, redistribution made only a small contribution to seed N. This result is in contrast with that of Day et al. (1979) using variety Minden on flinty-clay loam at Rothamsted, UK. When applied 6 weeks after sowing on 15 April 28.6% of ^{15}N was incorporated in the seeds, whereas only 7.7 - 8.4% of that added at pod initiation was recovered. This result could not be attributed to a lack of rainfall and the N-index is not known. These different results are not easy to explain and may be due to a number of factors such as the use of different varieties, different soil conditions and the outbreak of diseases and pests. This paper shows that differences in day and night temperature (thermoperiodism) play an important role (Figure 7). Similar results have been described by Went (1957). At low temperatures of 16°C leaves and pods senesce only very slowly and redistribution of N is most probably very small. Under such conditions the plants take up nitrate for a long period (Figure 7). The low redistribution value as found by Cooper et al. (1976) may have been caused by the low temperature during the growing season. This view is supported by the fact that the growing season was extremely long from May to November with full size but still green pods on September 9. It is well known that the beginning of senescence of the pods differs from year to year. In cool summers pods remain green for a

TABLE 2

SUMMARY OF PUBLISHED DATA ON SEED YIELD AND REDISTRIBUTION OF N IN *Vicia faba*

Author	Country	Var.	Yield kg/ha	% N-indices	% ^{15}N recovered in pods after application of $^{15}NO_3$ early September	Av. daily max. °C June - July - Aug.	Rainfall mm June - July - Aug.
Richards and Soper (1979)	Canada (Manitoba)	Diana	2 000	60	-	20.2 24.1 23.6	-
Dantuma and Klein Hulze (1979)	Netherlands (Wageningen)	Minica	5 780	90	-	← in 1978 → 19.5 19.9 19.7	125
Cooper et al. (1976)	United Kingdom (Long Ashton)	Maris B.	-	< 50	34	cool greenhouse	water applied as required
Day et al. (1979)	United Kingdom (Rothamsted)		4 000	-	8	unknown	210 (av. 182)
Sprent et al. (1977)	United Kingdom (Scotland)	Maris B.	6 712	59	-	← in 1974 → 16.7 19.1 18.4	144

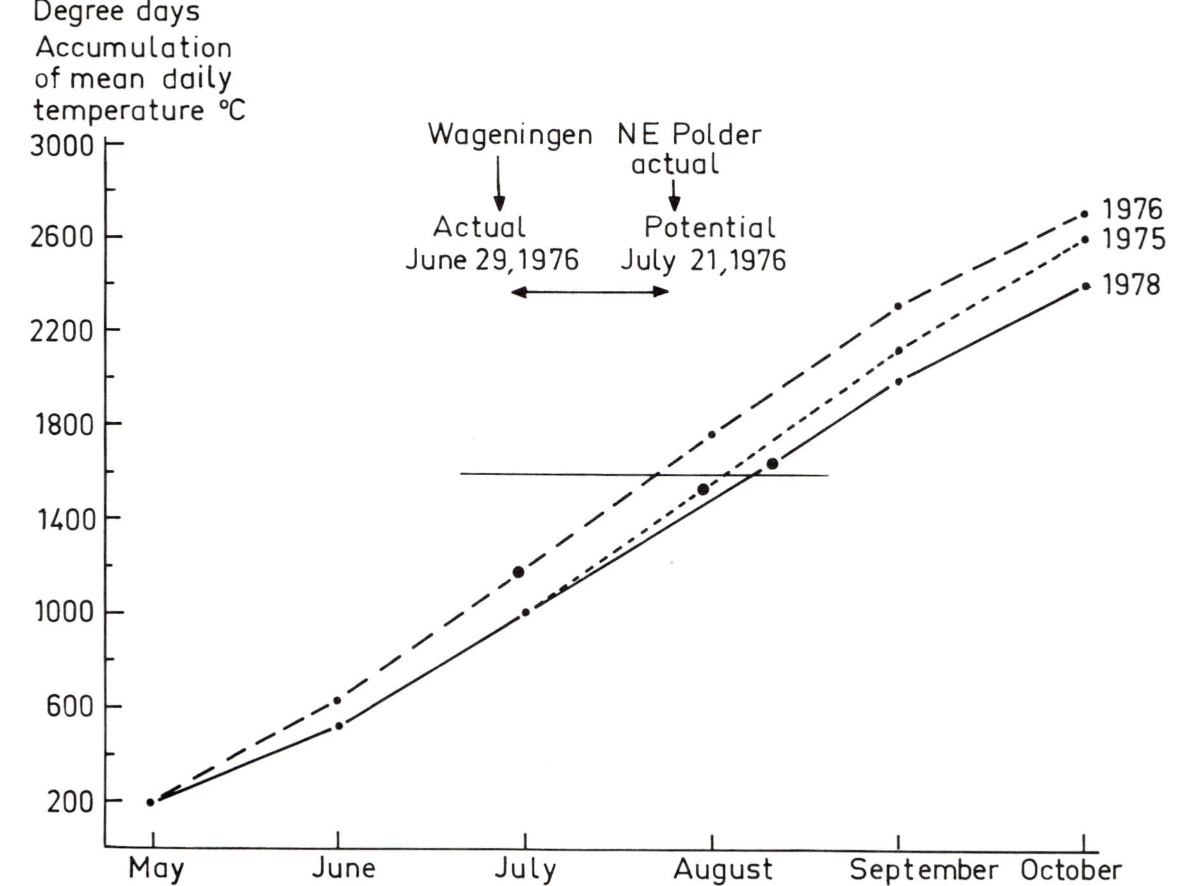

Fig. 10. Relation between maximal dry weight of leaves + stems (var. Minica) and accumulated temperature on a field plot (De Bouwing) near Wageningen. ● indicate dates at which plants reached their maximal dry weight of leaves and stems.

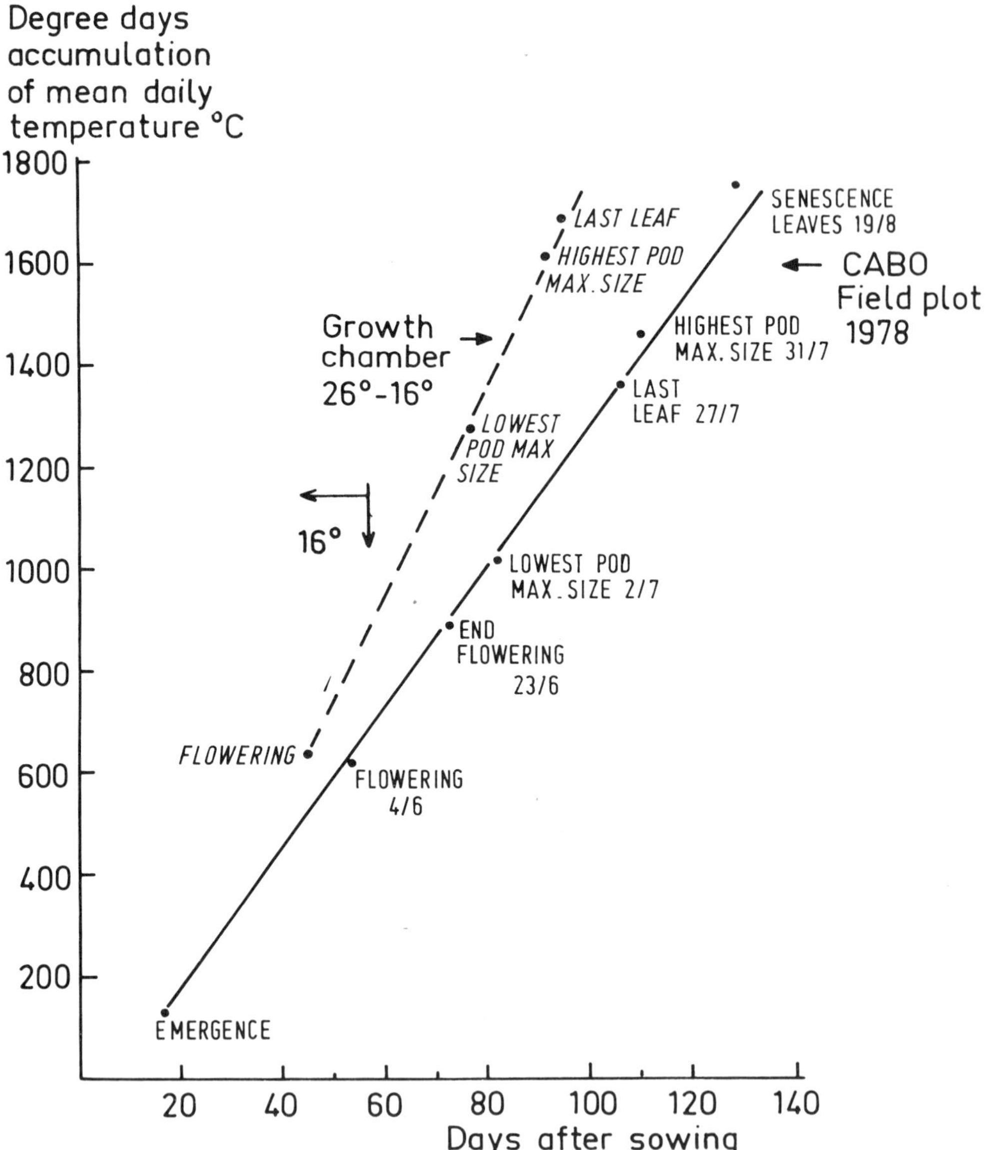

Fig. 11. Relation between developmental stages of var. Minica and accumulated temperature.

—— Indicate field plot at Wageningen (sandy soil). ---- Indicate experiment in a growth chamber. Conditions as in Figure 7.

long period and this affects the date on which the plants can be harvested (Hebblethwaite et al., 1977).

An attempt was made to use these kind of data to predict certain developmental growth stages. Data on accumulated average daily temperatures (^{o}C) were compared with dates at which stems + leaves of variety Minica reached their highest dry weight in the same field plot near Wageningen in 1975, 1976 and 1978. Values of 1 540 and 1 660°C were reached on July 28, 1975 and August 8, 1978, and at these dates stems + leaves reached their highest dry weight. Taking an average of 1 600°C, then plants should have terminated their growth on July 21 in 1976. Instead the highest dry weight was reached by June 29, 1976. This was certainly due to severe drought stress on this heavy clay soil with a high water retaining capacity. In fact, on a much better soil in the North East Polder the same variety terminated its growth at July 21, 1976 (Figure 10).

Figure 11 shows accumulated degrees plotted against developmental stages of a crop growing on sandy soil in comparison to plants grown in a growth chamber at 26°C during the day and 16°C during the night (conditions as in Figure 7). Plants flowered in both cases at about 600°C, but all other stages in the field experiment were reached at a lower value. It is known that a drought period occurred from June 12 to 23, 1978 and diseases and pests attacked the plants in the second half of July and in August. It does not seem unlikely that these stress conditions caused a premature breaking off of developmental stages. It may therefore be assumed that under non-stress conditions most stages would be reached at higher accumulated degrees in the field experiment (Figure 11).

These kind of data may be valuable to evaluate different developmental growth stages of the same variety in field trials in different localities in Europe. The system for predicting developmental stages will be more reliable if the available soil moisture and the water retaining capacity of the soil could be integrated in the system.

ACKNOWLEDGEMENTS

The authors are indebted to Mrs. ir. H.C.M. Chardon-Huisman for collecting data on accumulated daily temperatures and to ir. R. Booij for describing developmental stages of *Vicia faba* in a field study.

REFERENCES

Abdalla, M.M.F. and Fischbeck, G., 1978. Growth and fertility of stocks of field beans grown under three temperature regimes and the effect of natural water stress on seed index of a collection of *Vicia faba* l. Z. Acker und pflanzenbau (J. Agron. and Crop Science) 147, 81-91.

Cooper, D.R., Hill-Cottingham, D.G. and Lloyd-James, C.P., 1976. Absorption and redistribution of nitrogen during growth and development of field bean, *Vicia faba*. Physiol. Plant. 38, 313-318.

Dantuma, G. and Klein Hulze, J.A., 1979. Production and distribution in dry matter, and uptake, distribution and redistribution of nitrogen in *Vicia faba* major and minor. In: Some current research on *Vicia faba* in Western Europe. Eds. D.A. Bond, C.T. Scarascia-Mugnozza and M.H. Poulson. Commission of the European Communities, 369-406.

Day, J.M., Roughley, R.J. and Witty, J.F., 1979. The effect of planting density, inorganic nitrogen fertiliser and supplementary carbon dioxide on yield of *Vicia faba* l. J. Agric. Sci. Camb. 93, 629-633.

Gehriger, W. and Keller, E.R., 1979. Einfluss des Kopfens der Triebe auf die Versorgung der Bluten bei der Ackerbohne (*Vicia faba* L.) mit ^{14}C. Schweizerische landwirtschaftliche Forschung 18, 59-80.

Harper, J.E., 1974. Soil and symbiotic nitrogen requirements for optimum soyabean production. Crop Sci. 14, 255-260.

Hebblethwaite, P.D., Ingram, J., Scott, R.K. and Elliot, J., 1977. Some factors influencing yield variation of field beans *(Vicia faba)*. Scottish hort. Res. Inst. Bull. 15, 20-27.

Jones, R.W. and Sheard, R.W., 1977. Conditions affecting *in vivo* nitrate reductase activity in chlorophyllous tissues. Can. J. Bot. 55, 896-901.

Richards, J.E. and Soper, R.J., 1979. Effect of N fertiliser on yield, protein content, symbiotic N fixation in fababeans. Agron. J. 71, 807-811.

Said, H., Hegazy, T. and Iman, R.M., 1967. Growth, productivity and compositional contents of beans *(Vicia faba)* seeds as affected by different night-temperatures. Flora Abt. A. Bd. 569-576.

Sprent, J.I. and Bradford, A.M., 1977. Nitrogen fixation in field beans *(Vicia faba)* as affected by population density, shading and its relationship with soil moisture. J. agric. Sci. Camb. 88, 303-310.

Sprent, J.I., Bradford, A.M. and Norton, C., 1977. Seasonal growth patterns in field beans (*Vicia faba)* as affected by population density, shading and its relationship with soil moisture. J. Agric. Sci. Camb 88, 203-301.

Thibodeau, P.S. and Jaworski, E.G., 1975. Patterns of nitrogen utilisation in the soyabean. Planta 127, 133-147.

Thompson, R. and Taylor, H., 1979. Field plots for the practical estimation of potential yield. Scientia Hort. 10, 309-316.

Went, F.W. 1957. Experimental control of plant growth, Waltham, Mass. USA.

DISCUSSION

J.H.J. Spiertz *(Netherlands)*

I would like to ask Dr. Dekhuijzen about the results of his growth chamber studies. He shows two temperatures: one is a continuous temperature of 16°C, and the other is 26°C for the 14 hours of day and 16°C at night. There will be more rapid plant growth at the higher day temperature but the light level was kept the same for both temperature regimes. In the field there is a correlation between higher temperatures and increased radiation. Do you expect the same marked effect on the behaviour and growth of the plant when you extrapolate these results to the field?

H.M. Dekhuijzen *(Netherlands)*

It is very difficult to answer that question. I cannot increase the light intensity in the growth chamber. Different temperatures should now be examined, not only high day/low night, but also one temperature of an average of 22°C during the whole season, to see whether the plants then terminate growth prematurely. I know that it is extremely difficult to extrapolate data from the growth chamber to the field but it is an attempt, and I emphasise the word 'attempt'.

M.H. Poulsen *(Denmark)*

I understood you to say that only 25% of the total nitrogen uptake was by nitrogen fixation. Was that at a certain stage of the development of the plant or was it at the end of the season?

H.M. Dekhuijzen

No, this was during flowering. In that week, of the total N taken up, 25% was by nitrogen fixation and 75% by nitrate uptake. That is the only data we have from that field. It was not from the whole season unfortunately.

D.A. Lawes *(UK)*

That is far below the level I would have expected.

H.M. Dekhuijzen

It depends completely on the soil and the nitrate level of the soil.

J. Picard *(France)*

Have you any idea of what the genetic variability would be for the ability of the host plant for nitrogen fixation? In other words, do you think there are any differences due to the host plant?

H.M. Dekhuijzen

The capacity for nitrogen fixation? I doubt it but perhaps Dr. Lie has an answer? We worked together on the nitrogen fixation.

Lie *(Netherlands)*

I think there are some genetic traits which affect nitrogen fixation, but our experience with field beans is limited. I think the varieties used in our experiments are generally good nitrogen fixers. We have had a lot of experience with other legumes and we have had a number of varieties that differ in their capacity for nodulation and nitrogen fixation. There are many similar reports relating to clover, soyabeans, peas and peanuts.

D.A. Lawes

Substantiating that comment, we have found differences under sterile culture but we have not been able to reproduce these in the field.

E. von Kittlitz *(FRG)*

The problem with investigations on *Vicia faba* in growing chambers is that pollination and fertilisation may be inadequate.

G. Dantuma *(Netherlands)*

These experiments were conducted with inbred lines of c.v. Minica selected on auto-fertility.

FACTORS LIMITING GROWTH AND YIELD OF *Vicia faba*

R. Thompson and H. Taylor
Scottish Crop Research Institute,
Invergowrie, Dundee, UK.

ABSTRACT

Growing plants in highly fertile conditions (measured maximum MM yield plots) improved total dry matter production in 1978 and 1979, but yield of grain was better only in 1978. However, in two seasons of quite different weather conditions, growth and yield of Maris Bead in MM plots was very similar but differed markedly from the control plots. Total dry matter was also similar from MM plots and controls for each of three contrasting genotypes, Maris Bead, TI and CH421, but grain yield was higher from the controls which thus gave higher harvest indices.

Competition for light at several stages of growth was examined using cv. Herra grown in pots. Pots were changed from close to wide spacing or vice versa *for between 15 and 25 days and then returned to the original spacing for the rest of the season. The changes in spacing were made on four occasions, 21 June, 6 July, 31 July and 22 August. Marked effects on yield were found, for example 22 days at close spacing in otherwise wide-spaced plants halved the yield. Variation in the relative contribution to final yield of the various yield components was shown to account for the differences in yield.*

INTRODUCTION

Variability in yield of field beans between years or sites, as in other crops, is the cause of major planning problems for growers and processors. The variation results from edaphic and aerial environmental factors. Edaphic factors include soil physical structure, nutrition, water, pests and diseases, etc., all of which can potentially be controlled by the grower. The inclusion of water in this category may be questioned because, although the grower has some control over a shortfall, he has little control over an oversupply. The aerial environmental factors which affect growth and development include radiation, daylength and temperature. Radiation and temperature are significant, both in terms of the seasonal integrals and also possibly with respect to their marked short-term fluctuations. By providing super-fertile growing conditions and good pest and disease control (Thompson and Taylor, 1979), performance can be measured within the constraints imposed by aerial environmental factors only.

Clearly, an increased understanding is obtained if the broad partition of environmental effects between edaphic and aerial factors is complemented by detailed studies of responses to individual components such as radiation, water or temperature. The evidence accumulated for a variety of crops suggests that total dry matter production relates well to radiation integral. However, the partition of the dry matter between vegetative and reproductive growth in *Vicia faba* is particularly important. It is in this context that short-term fluctuations, in radiation for example, could have a marked effect on the development of individual yield components and subsequently on harvest index. Yield components develop sequentially, starting with flowering and ending with ovule growth, so it follows that stress at the end of the growing period could not affect pod set but could influence ovule size.

Experiments are described in which the growth and yield of several cultivars are examined when constraints due to edaphic factors are minimised to establish whether genotypes differ in their ability to exploit a favourable environment. Also, the effects were examined of short-term stress through competition for light at particular stages of plant development, to determine how short-term changes in canopy architecture (and hence intercepted radiation) influence yield through yield component interactions.

MATERIALS AND METHODS

Measured maximum (MM) yield experiments

A full description of the field technique used for examining growth with minimal stress is given by Thompson and Taylor (1979). The method consists essentially of growing high vigour seed, carefully planted and spaced, in an irrigated, highly fertile, synthetic compost, protecting the crop from pests and diseases, and providing support to prevent lodging. Crops of Maris Bead were grown by this method and compared with controls grown with good commercial husbandry, in 1978 and 1979. Also included in 1979 were a mutant with terminal inflorescence (TI) (obtained from D.A. Bond, PBI, Cambridge) and a Mediterranean landrace CH421 (obtained from D. Lawes, WPBS, Aberystwyth).

Competition for light

Effects of stress caused by competition for light were examined during different phases of development between flowering and pod-filling. Plants of the cv. Herra were grown in pots of UC compost (approx. 1 000 ml) placed at pot-spacings giving 5 or 100 plants/m^2 and shifted from one spacing to the other on one of four occasions between 21 June and 22 August. At the next change of spacing plants were either returned to their original spacing or left at their new one for the rest of the season. Details of the timing of the pot shifts are given in Table 1. The pots were watered by trickle irrigation and liquid feed was provided through the same system.

TABLE 1

TIMING AND DURATION OF CHANGES FROM WIDE TO CLOSE SPACING AND *vice versa*, AND SEED YIELD PER PLANT (g). CONTINUOUS LINE INDICATES TIME SPENT AT CLOSE SPACING AND DASHED LINE TIME AT WIDE SPACING

	Times of spacing changes						Wt. of seeds /plant g.
Date	20.3 (sown)	21.6	6.7	31.7	22.8	20.9 (harvest)	
Days	92	15	25	22	30		
Treatment							
Close spaced	———	———	———	———	———		7.0
Wide spaced	- - -	- - -	- - -	- - -	- - -		24.7
1	———	- - -	———	———	———		13.0
2	———	———	- - -	———	———		12.4
3	———	———	———	- - -	———		12.2
4	- - -	———	- - -	- - -	- - -		19.1
5	- - -	- - -	———	- - -	- - -		5.0
6	- - -	- - -	- - -	———	- - -		14.7
7	———	- - -	- - -	- - -	- - -		11.3
8	———	———	- - -	- - -	- - -		17.8
9	———	———	———	- - -	- - -		7.6
10	- - -	———	———	———	———		4.0
11	- - -	- - -	———	———	———		8.9
12	- - -	- - -	- - -	———	———		11.7

RESULTS AND DISCUSSION

Measured maximum (MM) yield experiments

Yields for Maris Bead in 1978 and 1979 (Table 2) show that total dry matter was greater and variation between years lower for the MM plot than for the control. However, whilst yield of grain was greater in the MM plot in one year, in the other year grain yield was less than in the control. Harvest index (Table 2) of this cultivar was clearly not favoured by highly fertile growing conditions, at least under the weather conditions of 1979. Components of yield for each crop are given in Figure 1, the length of each axis representing the value for each component indicated on the key; the intersection of the circle and the axis indicates the experiment average value for that component. Plant densities were similar for the four crops and the differences in yield per plant therefore

TABLE 2

YIELD (t/ha) AND HARVEST INDEX (HI) OF MARIS BEAD MM PLOTS AND CONTROLS IN 1978 AND 1979

	Total DM		Grain		HI	
	MM	Control	MM	Control	MM	Control
1978 Maris Bead	12.4	8.7	4.2	3.0	0.34	0.34
1979 Maris Bead	13.0	12.0	3.8	5.8	0.29	0.48

TABLE 3

YIELD (t/ha) AND HARVEST INDEX (HI) FOR MARIS BEAD, TI AND CH421 FOR MM PLOTS AND CONTROLS

	Total DM		Grain		HI	
	MM	Control	MM	Control	MM	Control
Maris Bead	13	12	3.8	5.8	0.29	0.18
TI	9.1	8.4	1.8	2.8	0.20	0.33
CH421	3.9	3.9	0.7	1.5	0.18	0.38

correspond closely to differences in yield per unit area. Much of the variation in yield between years for the control could be attributed to differences in pods/podding node, but values of this component were remarkably constant between years for the MM plot. Both beans/pod and average bean weight were consistent for all plots.

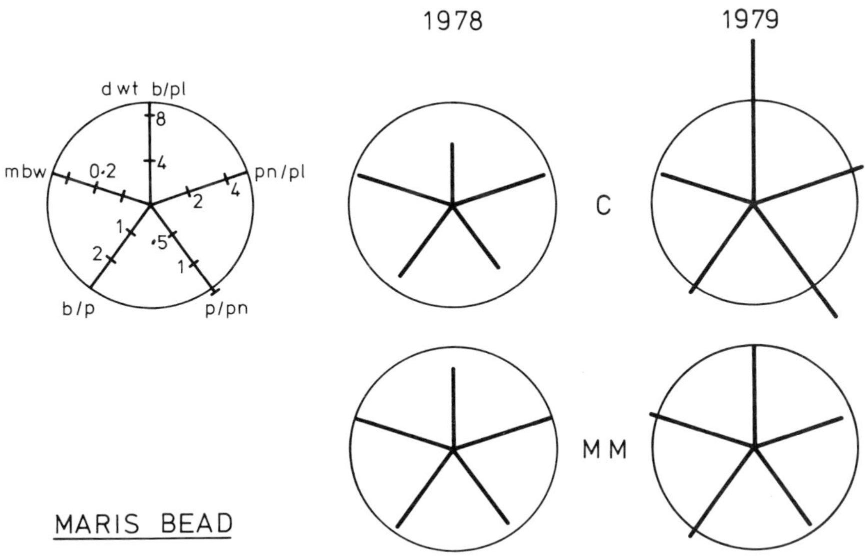

Fig. 1: Components of yield for Maris Bead from MM plots and control in 1978 and 1979 (d.wt. b/p = dry weight beans/plant, g. pn/pl = podding nodes/plant; p/pn = pods/podding node; b/p = beans/pod; mbw = mean bean weight, g.)

Comparing the three contrasting genotypes, total plant dry weights (Table 3) were similar to those of the MM plots and controls for each cultivar, and ranged from 3.9 t/ha for CH421 to 13 t/ha for Maris Bead. However, grain yields for each cultivar were lower for the MM plots than for the controls. In other words, the proportion of the total dry matter utilised for seed production (harvest index) was consistently lower from the plants grown at high fertility.

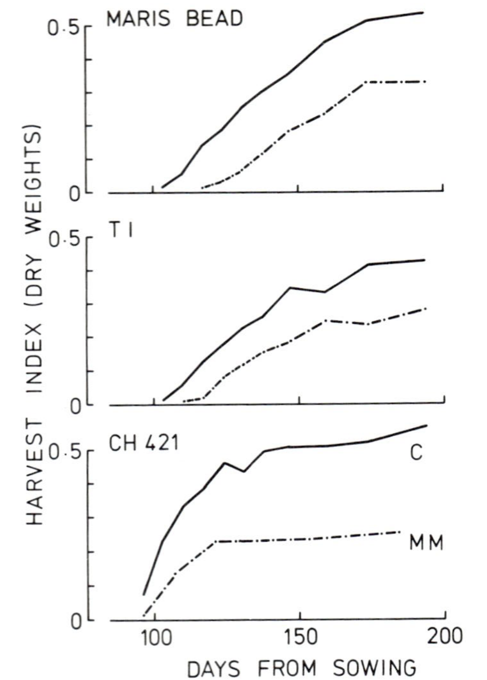

Fig. 2: Development of harvest index (dry weight of seed/total dry weight) of Maris Bead, TI and CH421 for MM plots and controls in relation to time.

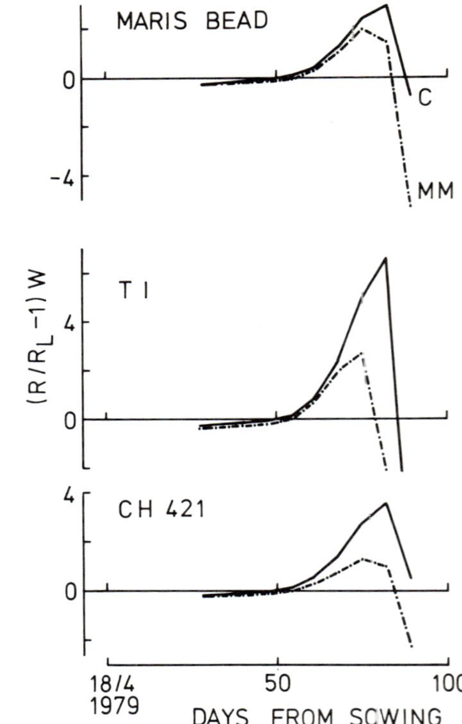

Fig. 3: Changes in value of W (R/Rl-1) with time for Maris Bead, TI and CH421 for MM plots and controls.

Throughout the period of pod filling (Figure 2) values in the ratio of seed to total plant weight were higher for the lower level of fertility, i.e. the control plots.

Whitehead and Myerscough (1963) suggested a method for estimating the levels of assimilate greater than that required to maintain a crop at its current level of vegetative growth and which would therefore be available for seed development. The surplus assimilate is defined as $S = W(\alpha - 1)$ (Figure 3) where $\alpha = R/R_L$ and where R = RGR and R_L = RLGR, and W = total dry matter (g). Trends for this parameter, in relation to time for the three genotypes, corresponded broadly with those of harvest index, but there is an indication that the vigorous indeterminate Maris Bead and the less vigorous but also indeterminate CH421 have rather less 'surplus' assimilate (S) than TI. It is tempting to interpret this difference as the result of less assimilate being needed for vegetative growth in TI, because of the topless habit but, as shown earlier, there was no evidence of improved yields from this genotype.

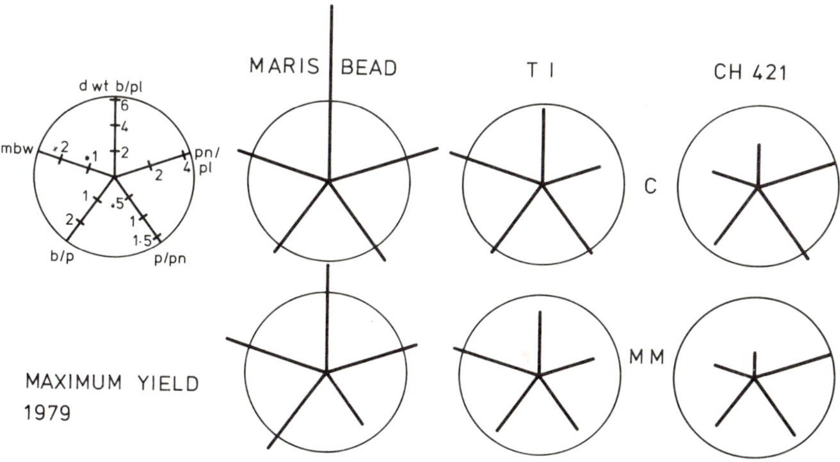

Fig. 4: Components of yield for Maris Bead, TI and CH 421 from MM and control plots. (See Figure 1 for key to symbols). Standard errors of the difference between two means with 2 d.f. were d.wt. b/p ± 1.93, pn/pl ± 0.84, p/pn ± 0.070, b/p ± 0.047, mbw ± 0.005.

The peak on each curve corresponds roughly to the time at which leaf area index reached a maximum. The large negative values of S which follow the steep decline in values, resulted mainly from losses of leaves and should be ingnored.

Highly fertile growing conditions promoted vegetative growth at the expense of reproductive growth, which suggests a way of introducing selection pressure for improved harvest index.

The components of yield of these three contrasting genotypes (Figure 4, circle = experiment mean) responded in broadly similar fashion to improved growing conditions, the reduced yields from the MM plots being accounted for particularly by reduced numbers of pods/podding node. Some changes in numbers of beans/pod occurred with TI and CH421 but not with Maris Bead, and may be of significance because this component has generally been considered to be relatively stable.

Thus there were marked differences in growth, development and yield between the MM plots and the controls, and it is now important to account for such differences in terms of the mechanisms involved, especially those controlling the partition of assimilates between vegetative and reproductive growth. The next section begins to explore one variable, light, in this respect.

Competition for light

Changing the level of competition for light at various stages of growth resulted in widely different yields (Table 1). As would be expected, the greatest contrast was between plants grown at each of the two densities continuously throughout the season, with three times the yield of seeds from the wide spaced compared with the close. Nevertheless, even a short period of time (about 15 days) at wide spacing for otherwise close spaced plants (Figure 5, no. 2) gave double the yield of the continuously close spaced plants. This suggests that an average

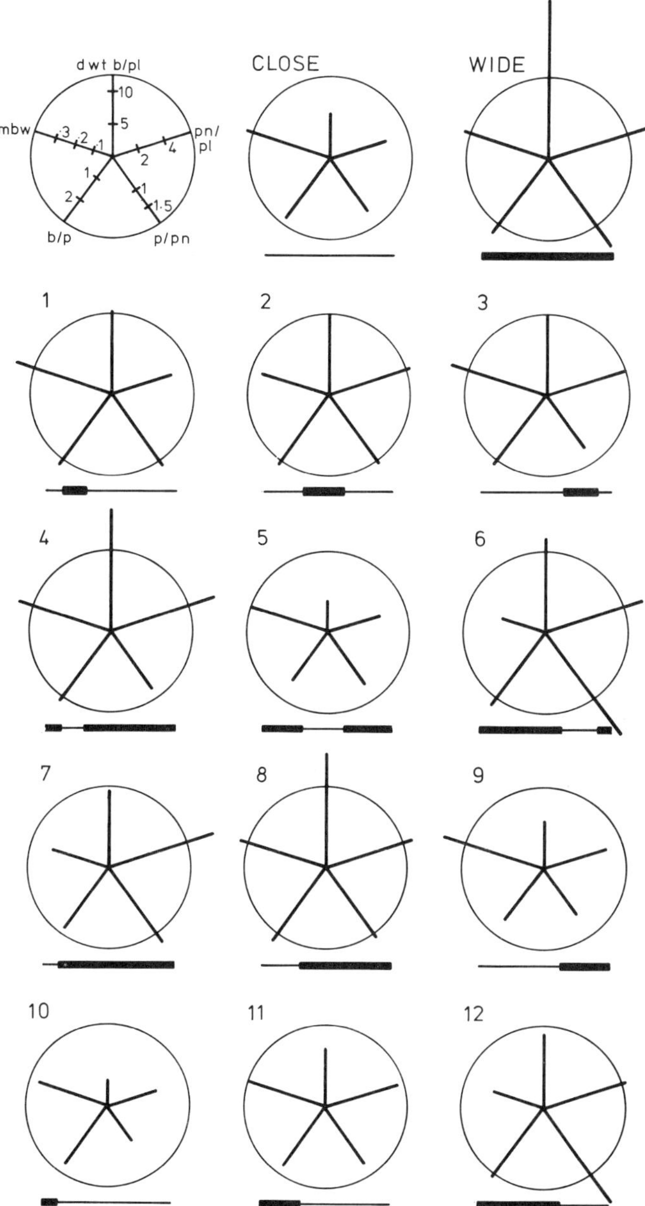

Fig. 5: Components of yield for cv. Herra when grown for varying proportions of the flowering and pod filling phase at two plant spacings. Duration at each spacing is indicated by the bar beneath each figure; wide bar ▬▬▬ = wide spacing, narrow bar ——— = close spacing, covering the period 21 June - 20 September. Standard errors of the difference between two means for 224 degrees of freedom were: d.wt. b/p ± 3.0, pn/pl ± 1.27, p/pn ± 0.25, b/p ± 0.26, mbw ± 0.039.

canopy may be able to support to maturity a considerably greater pod load than would normally develop, providing constraints to development are removed during a short but critical stage of growth. Conversely, when given a short period at close spacing and the rest of the time wide spaced, plants (Figure 5, for example no. 6) produced only about half the yield of the continuously wide spaced plants. It is also clear that the relative contribution of components of yield to final yield may vary (Figure 5) depending upon the stage of growth of the plants when stressed. Close spacing toward the end of pod development only (Figure 5, no. 6 and no. 12) reduced average seed size and, conversely, wide spacing at this time (Figure 5 no. 3 and no. 9) increased seed size. Competition stress in July during the period covering flowering and pod growth of the earliest set pods (Figure 5, no. 5) resulted in reduced numbers of pod-bearing nodes, pods/podding node and beans/pod, but had no effect on average seed weight, when compared with the continuously wide spaced.

Qualitative definition of the relationships between the several components of yield may enable the timing of stress in a crop of unknown history to be identified. Relationships between the patterns of yield component development and environmental variables would provide useful information to the plant breeder in search of components of yield known to be stable in respect of specific environmental variables.

ACKNOWLEDGEMENT

We thank Janet Brinklow for her very able technical help.

REFERENCES

Thompson, R. and Taylor, H., 1979. Field plots for the practical estimation of potential yield. Scientia Horticulturae, 10, 309-316

Whitehead, F.H. and Myerscough, P.J., 1962. Growth analysis of plants. The ratio of mean relative growth rate and leaf area increase. New Phytologist, 61, 314-321.

DISCUSSION

C.L. Hedley *(UK)*

Did you do your maximum yield experiments at single planting density?

R. Thompson *(UK)*

Yes, generally, although I have some data from one experiment with a range of different densities.

C.L. Hedley

With the sort of differences in harvest index that you found, would you not expect that if you gave a plant conditions where its growth rate would increase, you would increase intra-plant competition? Then you would expect a fall-off in yield; for example, in high fertility conditions the plant would behave as if it was a high planting density.

R. Thompson

Yes, lower plant densities give higher harvest indices than higher plant densities.

C.L. Hedley

When you increase the fertility the plant produced will mimic that from a high planting density?

R. Thompson

It could be. Therefore are plant breeders right to make selections at low density?

G.P. Chapman *(UK)*

I wanted to follow up an invitation which I have made to Dr. Thompson to come to Wye, because it is an appropriate time to report on an experiment being done by my colleague, Gerry Goldwyn. He has been trying to do 10 years' work in one - I mean that quite seriously. He has taken two varieties of field beans and planted them at half-weekly intervals since the start of the growing season. Then a fairly ambitious computer correlation is made between yield performance and the various ingredients in the environment. What he has been doing is a correlation for time intervals of from 1 - 10, 2 - 11, etc., for temperature, soil moisture and such with growth. Perhaps this is a slightly different way of examining the problems that you have reported but I would think that you both would find it of mutual benefit to look at this particular technique.

R. Thompson

Thank you for the invitation, Dr. Chapman, I would like to come. I think that this is a good idea and we have been using the technique with the vegetable crop calabrese. Sowing dates are used to give us different radiation levels etc. One of the problems is that so many factors interact and are confounded that it takes a fairly powerful programme or model to sort out exactly what relationships exist.

D.A. Lawes *(UK)*

In your maximum yield plots, were you not concerned about the effect of the bird netting over the plots? This must have had some effect.

R. Thompson

We measured the light cut-out caused by the netting and it was about 5%; we did not think that mattered too much. It also reduced wind, but we erect wind shelters around the plots anyway.

However, we did have a catastrophe with one net because its mesh was so small (40 mm) that snow collected on it and caused the whole structure to collapse. So we are now using a 100 mm mesh material, the idea being that this will keep out the larger vermin, pigeons are a particular problem with field beans, and we have wire netting round the bottom to keep out four-legged vermin.

D.A. Lawes

Thank you very much, Dr. Thompson.

DROUGHT TOLERANCE, DEFINITION AND MEASUREMENT

A.F. van der Wal
Foundation for Agricultural Plant Breeding,
Wageningen, The Netherlands.

ABSTRACT

The various injurious effects of water stress on crop plants are discussed in relation to stages of growth and development and to crop loss. Plant breeders can use this knowledge to develop screening methods in the various regions where certain effects of stress are of critical importance.

Examples of some screening procedures used at the Foundation for Agricultural Plant Breeding are given and include screening in the greenhouse of young plants; yield tests on large lysimeters and soil depth gradients.

INTRODUCTION

The boom in farm yields after World War II until about 1970 is mainly attributed to massive inputs of energy in agriculture by means of fertilisers, pesticides, irrigation and mechanisation, as well as to genetic improvement of the crops. Since 1970, the yield increases per unit area have levelled off (Karamanos, 1979), and the continuously increasing costs of energy and capital are forcing a gradual change in research priorities. A more intensive study of the factors limiting crop production is necessary. Water is a well-known key factor in plant growth, and numerous studies have been done to understand its mode of action (cf. Hsiao, 1973). A short description of some aspects of water stress for the growing and developing plant are given here, as well as a breeder's approach to improving the physiological potential of crop plants under water stress.

EFFECTS OF WATER STRESS ON GROWTH AND DEVELOPMENT

The injurious effects of low soil water potentials on plant growth and production vary with the stage of plant development at which stress occurs. Water shortage during the first weeks of growth causes slower leaf expansion and stem elongation of the young plants compared to plants grown at optimal soil water potentials. These plants obviously develop a more xerophytic habit and, together with a reduced growth rate, development can also be delayed. Although delay of growth and development does not necessarily result in crop loss, a delayed crop is to be considered a high risk crop, because of lost time. Adverse conditions at the end of the growing period such as drought, heat, frost, pathogens and pests, may limit the potentials of such delayed crops. Cell division is relatively unaffected at moderate stresses, while leaf expansion and stem elongation are suppressed to a considerable extent. Plants gradually adapting to decreasing soil water potentials seem relatively insensitive to wilting.

A dry period, some 6 - 8 weeks after emergence, causes a reduction of the transpiration rate and, to a lesser extent, a reduction of the photosynthetic rate. Under field conditions, especially at high light intensities, lowering of soil water potentials affects the transpiration rate initially much more than the rate of photosynthesis.

Tillering of cereals may be impaired, and a prolonged period of drought causes wilting, chlorosis and necrosis of the leaves. The ability to recover from a period of severe water stress seems a valuable property of plants in a crop situation.

Flowering and seed setting time is a critical period for many crops. Any disturbance of the plant's internal water potentials by either drought or disease, or a combination of both, can cause abortion of newly formed seed (van der Wal, 1975). A serious reduction in seed number is usually only compensated for in part by a higher average seed weight. Seed abortion is therefore serious and usually results in crop loss.

During the period of seed filling, the daily dry matter production and the length of the period of crop growth determine the yield. Water shortages in this period reduce the photosynthetic rate, and prolonged periods of water stress enhance senescence and premature loss of foliage. The injurious effects of water stress are functions of duration of drought, the daily potential transpiration of the crop and the stage of development.

The photosynthetic efficiency of a stressed crop can be affected in four ways:

1. By reduction of the photosynthetic rate due to closure of the stomata, and possibly later by direct effects of low leaf water potentials.
2. By reduction of leaf expansion during the early stages of growth resulting in a lower degree of light interception.

3. By reduction of the 'sink' size, causing a lower photosynthetic rate via a feedback control system.

4. By the irreversible loss of leaves by chlorosis and necrosis after a prolonged period of stress.

BREEDING TO IMPROVE THE PHYSIOLOGICAL POTENTIAL OF STRESSED CROPS

Using Levitt's (1972) terminology, drought resistance can be divided into drought avoidance and drought tolerance.

Drought avoidance is, by definition, the ability of the plant to maintain a high internal water potential when exposed to environmental drought. Improvements can be obtained either by increasing the plant's means of extracting water from the environment, e.g. by extending the root system, or by saving water by stomatal control, increased cuticular resistance and reduction of transpiration area. Another way is the 'true drought tolerance'; which is the ability of the plant to endure low internal water potentials, e.g. by dehydration tolerance or by osmoregulation.

For our climatic conditions, it seems worthwhile to screen for genetic variation in response to water stress of plants 6 - 8 weeks old. Large numbers of plants can be grown at optimal soil water potential and after 6 - 8 weeks, depending on the stage of development, water is withheld until the transpiration rate, monitored by means of the sap flow meter, is almost zero for one week. Plant responses, such as wilting, chlorosis, necrosis and recovery after rewatering are scored. Large variation among potato cultivars has already been found, and tests with field beans are now in progress.

A physiological explanation of these phenomena during decreasing soil water potentials and the recovery from stress may well be a genetic difference in stomatal control, and thus in transpiration pattern, resulting in a more or less rapid consumption of the available water in the soil (drought avoidance). Another explanation may be that the internal water potentials become very low, but the plants are really tolerant and endure the stress without symptoms such as chlorosis and necrosis.

A further point of interest is the possibility of screening for genetic differences in the transpiration pattern of cultivars and lines. It is now possible to monitor the transpiration patterns of many genotypes simultaneously by means of the sap flow meter.

Relatively large lysimeters are used to study drought in the field. Soil water potential can be controlled and effects of drought on yield can be studied with stress at various stages of plant development.

In another type of field experiment, a gradient for rooting depth is used. The depth of the bin buried in the soil varies from 25 to 80 cm over a length of c. 10 m. Various genotypes are planted in rows spanning the range in rooting depth. Because of the difference in water availability, due to differences in soil volume, symptoms of drought appear first at the 25 cm and gradually progress to the deep end of the bin. Yield differences in response to this gradient indicate the degree of drought tolerance of the genotype. Yield trials, however, can only be made on limited numbers of genotypes.

REFERENCES

Hsiao, T.C., 1973. Plant responses to water stress. A. Rev. Pl. Physiol., 24, 519-570.

Karamanos, A.J., 1979. Water stress: a challenge for the future of agriculture. In: Plant Regulation and World Agriculture, T.K. Scott (Ed.) Plenum Press, New York - London, 415-455.

Levitt, J., 1972. Responses of Plants to Environmental Stresses, Academic Press, New york - London, p.697.

van der Wal, A.F., 1975. An ecophysiological approach to crop losses, III: Effects of soil water potential on development growth, transpiration, symptoms and spore production of leaf rust infected wheat. Neth. J. Pl. Path. 81, 1-13.

DISCUSSION

H.M. Dekhuijzen *(Netherlands)*

I cannot understand your remark about the relationship between disease and water stress. With water, did you not see any effect on the nematodes?

A.J. van der Wal *(Netherlands)*

No, these were both field plots. There was no control of water potential at all. The only difference was that the poor one was for the summer of 1976 and it was considered to be a dry plot. The other was for the summer of 1979 and that was a plot watered daily; we had showers every other day. That was the effect on an interaction of keeping the soil water potential at a very high level.

G. Pommer *(FRG)*

Have you found significant differences between varieties with your method of measuring transpiration rates, and is there any connection between these differences and the yield under dry conditions?

A.F. van der Wal

The sap-flow meter is only in the development stage. We have run this machine for about three years now, under field conditions, and we have found indications of genetic differences in the transpiration pattern. However, it has not been tested for a whole period or with many genotypes; in principle this can be done.

G. Pommer

Did you find any connection with yield under conditions of drought?

A.F. van der Wal

No, I did not try. When we were testing these sensors, we measured the patterns only for a few days and not for the whole cycle.

G. Pommer

Will you be presenting a paper on this method when you are sure it works?

A.F. van der Wal

There are already two papers on this method. The first, with similar results to those given here, was published in 1978. Last year the technical side of it was published.

H.M. Dekhuijzen

My question is related to what Dr. Pommer has said. How do you relate differences between genotypes with all the transpiration data?

A.F. van der Wal

The impression is quite clear and I think we can generalise. Usually the transpiration rate has a lower effect than photosynthesis. If we wish to find 'water savers' in terms of drought resistance, meaning those plants which use a minimum of water and give an optimal yield, then we should first of all look for differences in the transpiration pattern. When we find these differences, under certain standardised conditions, then we must find an explanation for these differences. Is this stomatal control? Has it to do with root systems? However, first of all we need a screening method for genotypes for the differences in the transpiration pattern, and this is our aim for the next three or four years.

H.M. Dekhuijzen

I believe there is already a relationship for the potato, between different genotypes and transpiration.

A.F. van der Wal

We have found some varietal differences in transpiration patterns.

M.C. Saxena *(Syria)*

There are other methods for measuring the water potential of the plant and transpiration as such. Do you think that the sap-flow meter is preferable to the others, for example, diffusive resistance measurement, or measurement of leaf water potential as such by alternative methods?

A.F. van der Wal

The leaf water potentials can be measured quite easily, by a pressure bomb, for example, and can be standardised by making the measurement at 4 o'clock in the afternoon, which is a standard procedure, and then you will get reasonable impressions. Unfortunately, I am very lazy and I prefer to have an automatic system, which operates in a non-destructive way for many genotypes. The output of this machinery is about one millivolt; it can be recorded and put into an automatic data processing system so easily that I prefer it to a destructive one which uses much plant material.

M.C. Saxena

Do you take the measurement at a specific time during the day, or do you have continuous recordings?

A.F. van der Wal

At the moment we are labour saving because the system is connected to a recorder; we get a mass of data. But we want to get a complete picture of daily curves. We have done this for a few weeks but not for a whole season on a continuous basis.

M.C. Saxena

With the last method that you showed, relating to the rooting depth and the soil depth gradient, do you think there is any lateral movement of water that might affect the water used by the shallower rooting system in contrast to the deeper one?

A.F. van der Wal

I did not find any indications that it is a serious problem but I cannot prove that it is not. The dimensions of such a plot means that even in the shallower parts of 25 cm , more than 50 l of soil are still available per plant; that is quite a lot. It is more than many plants in many countries have in field conditions. Therefore I do not think it is a great problem.

D.A. Lawes *(UK)*

One of the problems, from the breeding point of view, is to pick out certain characters which are related to drought tolerance. One thing we have been doing is to look at stomatal size and frequency. In fact, we have found genotypic differences in water use and growth with different types grown under different water regimes. The difficulty is in trying to exploit that in terms of varietal production; it is a tremendous step to have to take. With your technique it is a question of which types are shown to be better than others in retrospect; whether or not you can actually initiate breeding programmes from this data, I am not sure. Would you care to comment on that? Would this type of work allow initiation of breeding programmes?

A.F. van der Wal

For screening populations, we should have a technique other than just the sap-flow meter. Even if it were possible to screen about 100 genotypes at one time, it is still only 100 genotypes and not 10 000, which is the order of magnitude for screening populations. We feel that we should screen the large quantities using relatively young plants looking at chlorosis and necrosis recovery as a first step. Then we could use this technique as the next step to explain what goes on, what causes a plant to wilt at a certain stage or at a certain speed, or what makes it recover? Then there is the leaf water potential measurements, stomatal densities and frequencies, then transpiration and even root studies, although this is not for massive screening - only for cultivars, varieties and similar levels.

A FIELD STUDY OF THE CARBON ECONOMY OF NORMAL AND 'TOPLESS' FIELD BEANS *(Vicia faba)*

R.B. Austin, C.L. Morgan and M.A. Ford
Plant Breeding Institute,
Cambridge, UK.

ABSTRACT

The growth and yield of Maris Bead spring bean and Topless, a population incorporating the terminal inflorescence mutant, were compared in irrigated and non-irrigated experiments at Cambridge in 1979. In both experiments, Topless, initially grew more rapidly than Maris Bead, but there was little difference between the two genotypes in total dry matter yield at maturity. The yield of grain from Topless was 89% of that of Maris Bead, in both experiments. Irrigation, given in one experiment until the onset of pod growth, greatly stimulated the vegetative growth of both genotypes but increased their grain yields only slightly. At maturity, Topless plants were shorter than those of Maris Bead, particularly in the irrigated experiment. The yield of vegetative dry matter in Topless depended on the production of tillers and axillary branches, many of which did not bear grain and were sinks for assimilate during grain filling.

Patterns of photosynthesis and translocation were studied by ^{14}C *methods. Until the terminal inflorescence in Topless was produced and pod growth began, the patterns of assimilation by both genotypes were similar. During pod growth, the leaves contributed most to the photosynthesis of the canopy, and the large, upper leaves were the main source of assimilates in Topless. In Maris Bead more leaves contributed assimilate and no single leaf was a dominant source. Pods contributed a maximum of 20% to the gross assimilation of the crop. The rate of photosynthesis per unit dry weight of pod, judged from* ^{14}C *fixation, was greater in Topless than in Maris Bead.*

It is concluded that the 'topless' plant type is not inherently inferior in productivity to the normal indeterminate type. It is suggested that in future breeding to produce determinant varieties having the terminal inflorescence character, selection should be practised for increased production of tillers which develop synchronously with the main shoot and against the production of infertile branches.

INTRODUCTION

The recognition by Sjödin (1971) of a mutant of field beans giving a terminal inflorescence has excited the interest of breeders and physiologists as it appeared to offer the possibility of a built-in control of 'excessive' vegetative growth. 'Excessive' vegetative growth occurs in wet years on heavy soils and is believed to reduce grain yield by causing a diversion of assimilates which would otherwise be available for floral development and grain growth. It also makes the plants susceptible to lodging, making harvesting difficult. A modest breeding programme to incorporate the mutant into a suitable background has been started at the Plant Breeding Institute and the aim of the study reported in outline here was to provide an early indication of the value of the mutant in the breeding for high and stable yield.

MATERIAL AND METHODS

The terminal inflorescence mutant originally obtained by Sjödin (1971) was backcrossed to two PBI inbred lines and after selecting for suitable agronomic type a new population referred to here as Topless, was formed.

Two adjacent experiments, each with four replicates of plots of Topless and Maris Bead, were sown on 14 April 1979. Seed was sown in rows 30 cm apart to give a nominal plant density of 30 plants/m^2. Plots were each 4 rows x 15 m long (19 m long in two replicates) and each experiment was surrounded by guard strips 4 rows wide. One experiment was irrigated frequently for six weeks from the beginning of June to maintain the soil at or near field capacity. Irrigation was stopped in July because of the risk of lodging. The other experiment received only natural rainfall (192 mm between sowing and harvest).

To measure dry matter production, plants from 4 x 50 cm lengths of row were harvested fortnightly from each plot on six occasions (seven on the irrigated experiment) and the dry weights of the vegetative parts and, when present, of the pods and seeds were determined. The last sample was taken at maturity and provided an estimate of yield and harvest index.

To determine the contributions to the photosynthesis of the canopy made by component organs a ^{14}C method was used. It was essentially the same as that described by Austin et al. (1976). In two plots of each genotype in each experiment four metal bases were inserted into the ground shortly after emergence of the seedlings. These were used later to accommodate the plastic cuvettes which covered the plants in a 60 x 60 cm cropped area when they were labelled with $^{14}CO_2$. Labelling was carried out on four occasions. The ^{14}C in the different organs of the plants harvested immediately after labelling, before any significant translocation had occurred, was taken as a measure of their rates of gross photosynthesis.

To trace the fate of the carbon assimilated by individual leaves in the canopy, leaves were enclosed in a polyethylene bag and supplied with 10 μCi $^{14}CO_2$ in air. They were allowed to assimilate for 10 min following which the bags were removed. Plants so labelled were later harvested and the ^{14}C content of the component organs, including the tap roots, determined.

RESULTS

Growth and yield

Topless plants were initially somewhat taller than those of Maris Bead but the general appearance of the genotypes was otherwise similar until the terminal inflorescence was produced in Topless in late June. Following this, the irrigated Maris Bead continued to increase in height, reaching 162 cm at maturity. Irrigated plants of Topless were 94 cm tall at maturity. On the non-irrigated plots, Maris Bead was 111 cm and Topless 90 cm tall at maturity.

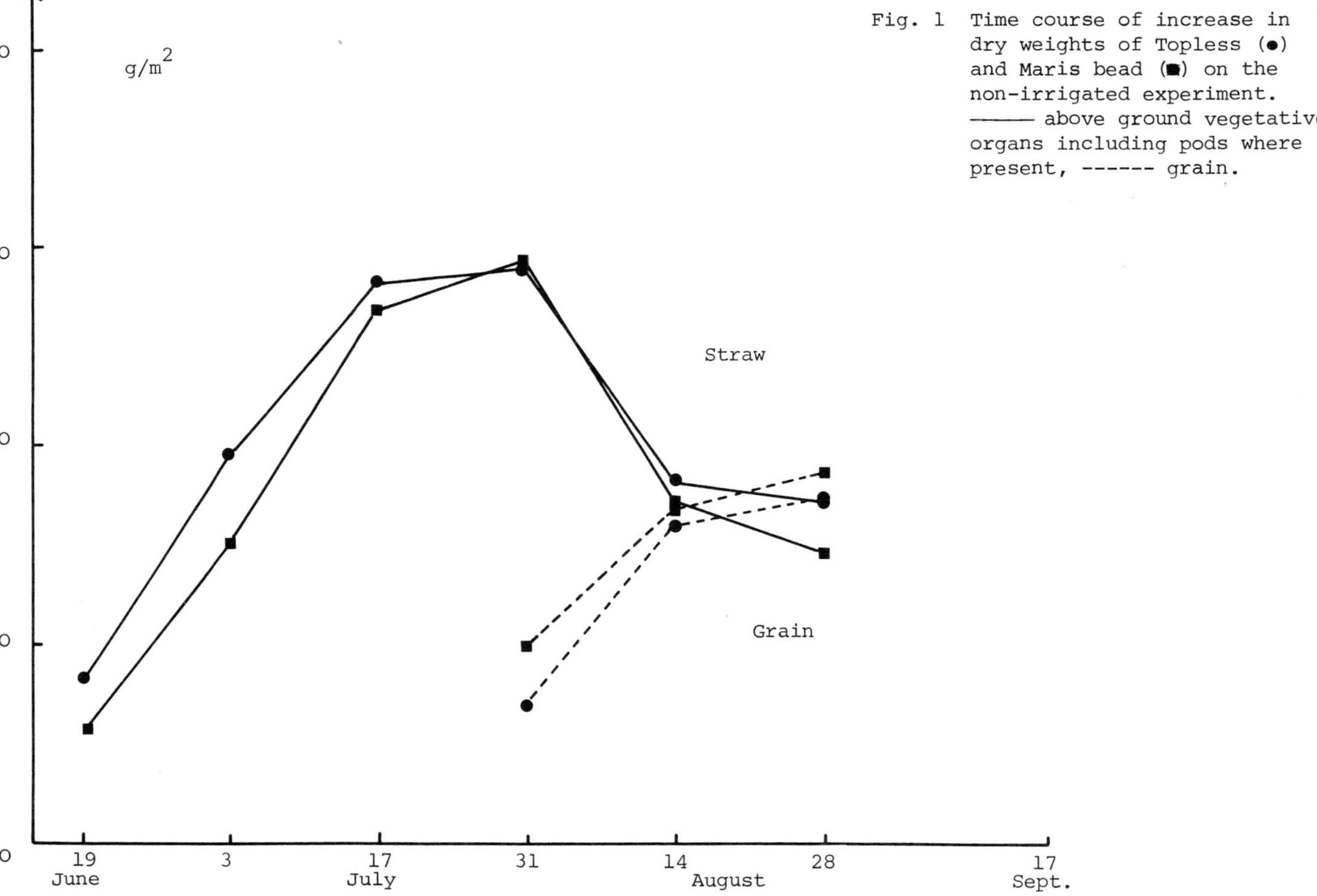

Fig. 1 Time course of increase in dry weights of Topless (●) and Maris bead (■) on the non-irrigated experiment. ——— above ground vegetative organs including pods where present, ------ grain.

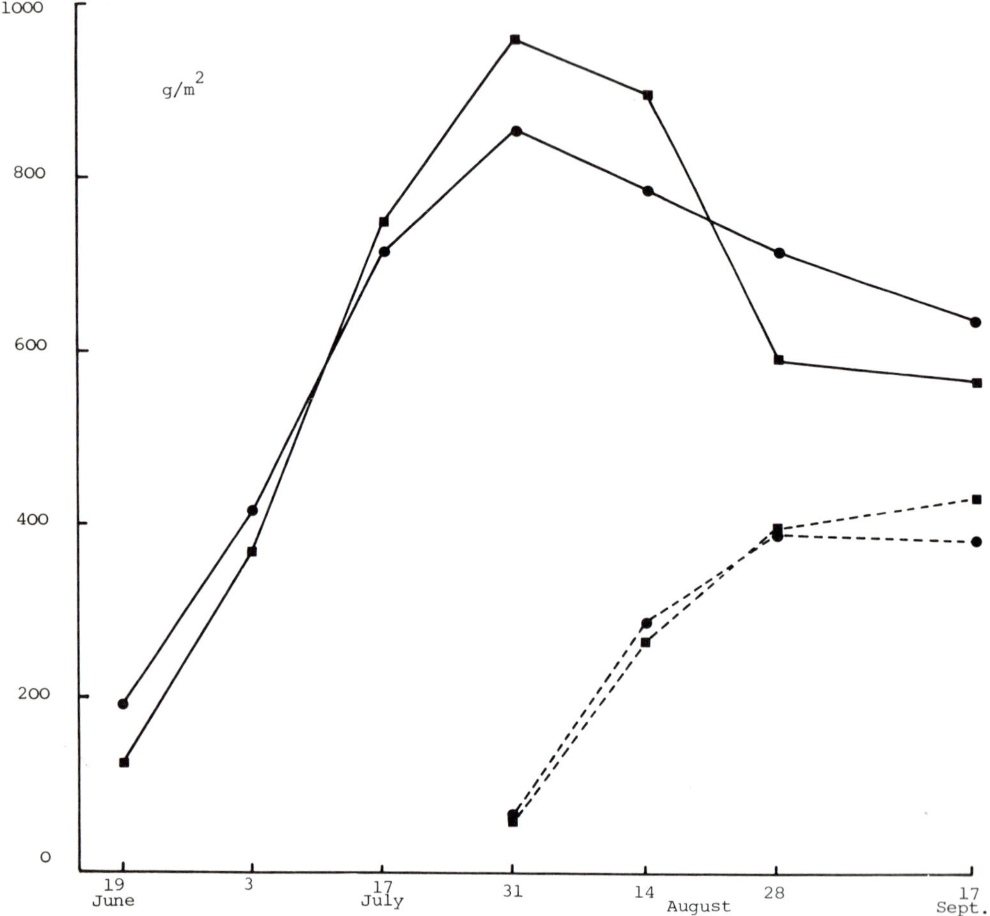

Fig. 2 Time course of increase in dry weight of Topless and Maris Bead on the irrigated experiment. Symbols as for Figure 1.

The time course of change in dry weight of the above-ground plant parts is shown in Figures 1 and 2. In both irrigated and non-irrigated experiments Topless plants were heavier than those of Maris Bead at the time of the first harvest on 19 June. This difference persisted until 17 July. After the onset of grain growth there was a loss in dry weight from the vegetative organs of both genotypes amounting to around 200 - 300 g/m^2 (50 and 30% respectively of their maximum dry weight in the non-irrigated and irrigated experiments). This loss in dry weight was considerably greater than that expected from the loss of leaf laminae. On 2 August, when most leaf laminae were still attached to the plants, they constituted approximately 13% of the total above ground plant dry weight. At maturity, in both experiments, the weight of vegetative dry matter was greater in Topless than in Maris Bead.

Irrigation almost doubled vegetative dry matter production but it had much less effect on grain yield, increasing it when averaged over varieties from 357 to 409 g/m^2 (Table 1). Although Topless yielded around 10% less grain than Maris Bead in both experiments, the difference was not large enough to reach statistical significance, and there was no indication that the yield of Topless was increased more than that of Maris Bead by irrigation.

Contributions of various organs to canopy photosynthesis

The uppermost fully expanded leaf, generally the largest on a plant, was used as a common reference point. Leaves above it were numbered +1, +2 etc. and those below it -1, -2 etc. The term 'apex' is used here to mean all leaves and the stem within the cluster of unexpanded leaves not individually exposed to the light.

On 22 June, before the emergence of the terminal inflorescence in Topless, and when the plants appeared similar in all plots, the patterns of photosynthesis in the canopies were similar. The lamina of the uppermost fully expanded leaf, and the one above and below it were responsible for almost half the photosynthesis of the entire canopy. Stems and petioles together contributed 15%.

TABLE 1

GRAIN YIELDS AND THEIR COMPONENTS, AND HARVEST INDEX

	Yield of dry grain g/m^2	Mean seed weight mg	No. of seeds per pod	No. of pods per m^2	No. of plants per m^2	No. of fertile basal branches per m^2 †	Harvest index %*
Not irrigated							
Topless	345	478	2.68	269	33.3	24.7	50
Maris Bead	370	336	3.39	325	29.0	4.3	56
Irrigated							
Topless	385	419	2.55	359	32.3	30.0	38
Maris Bead	433	328	3.48	379	27.7	3.0	43

*Yield of dry grain as a percentage of the total above-ground dry matter at maturity

† excludes main shoot.

Results of the labelling on 2 August are presented as being typical of the pattern of assimilation in the canopy during grain filling (Figure 3a-d). In Topless in both the irrigated and non-irrigated experiments, the uppermost leaf or the one below it contributed the most to canopy photosynthesis, other leaves contributing progressively less. As on 22 June, stems and petioles contributed about 15%. Photosynthesis in each pod of Topless was generally less than a quarter of that of the subtending leaf. Branching* was very variable in this

*Branches are defined here as infertile shoots. They usually develop as from the axils of leaves, but sometimes from the main stem at ground level. Tillers are defined as fertile shoots developing from the main stem at or near ground level.

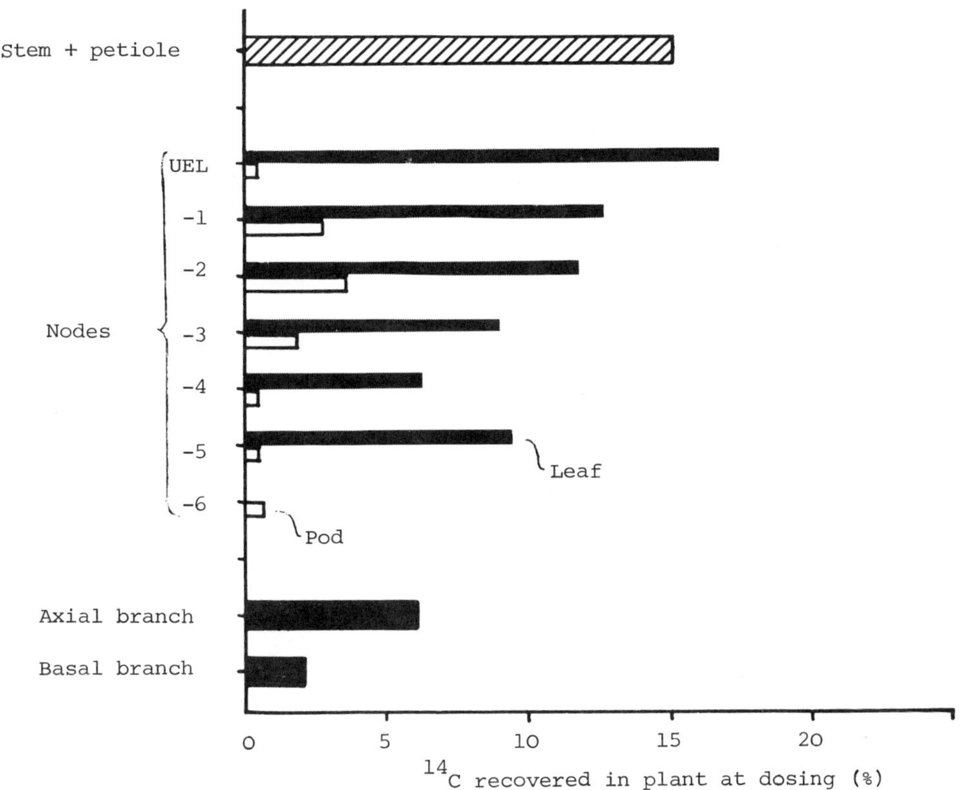

Fig. 3a Percentage contributions of component organs to the photosynthesis of the entire canopy on 2nd August. UEL = uppermost expanded leaf. Topless, non-irrigated.

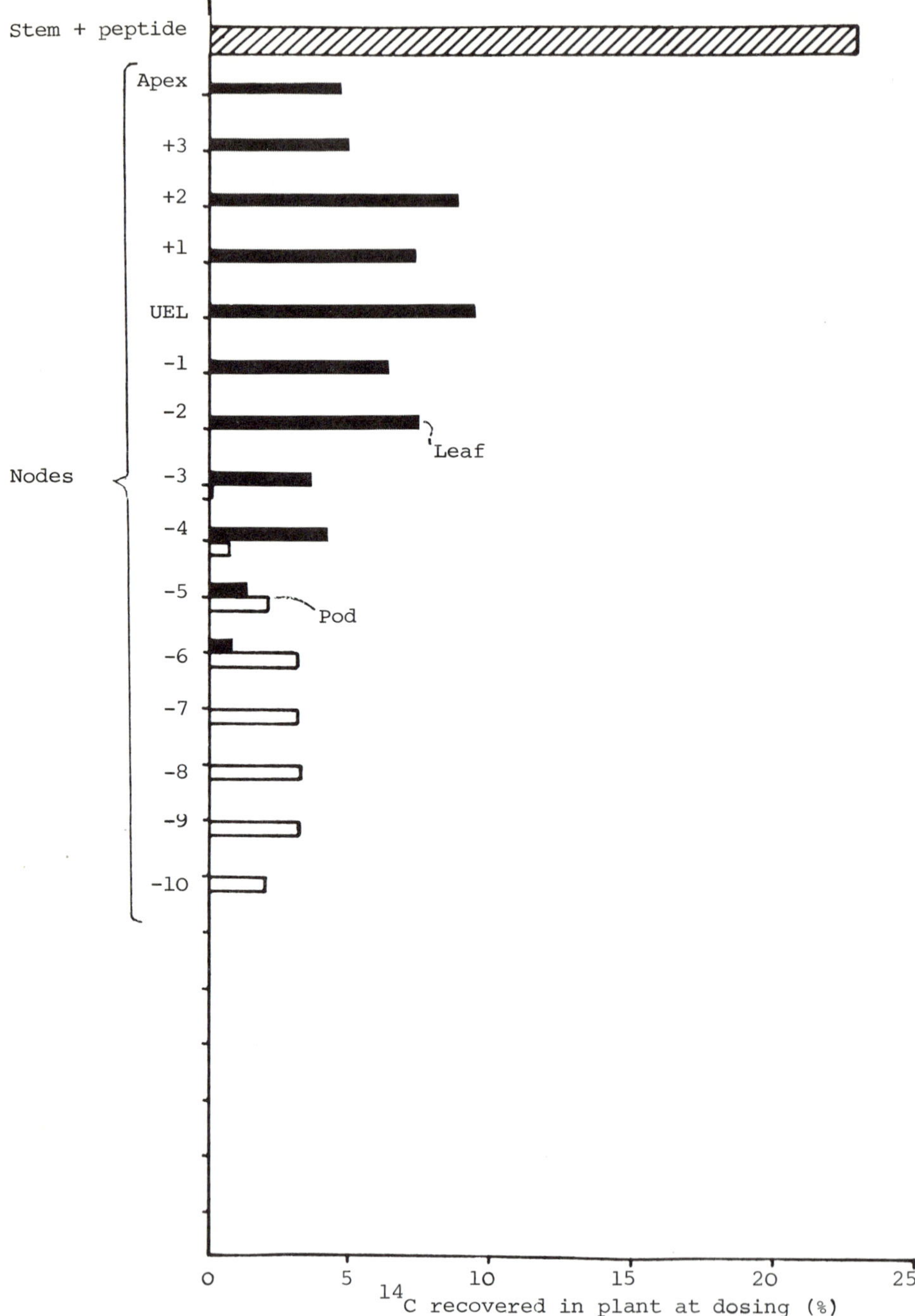

Fig. 3b Percentage contributions of component organs to the photosynthesis of the entire canopy on 2nd August.
Maris Bead, non-irrigated.

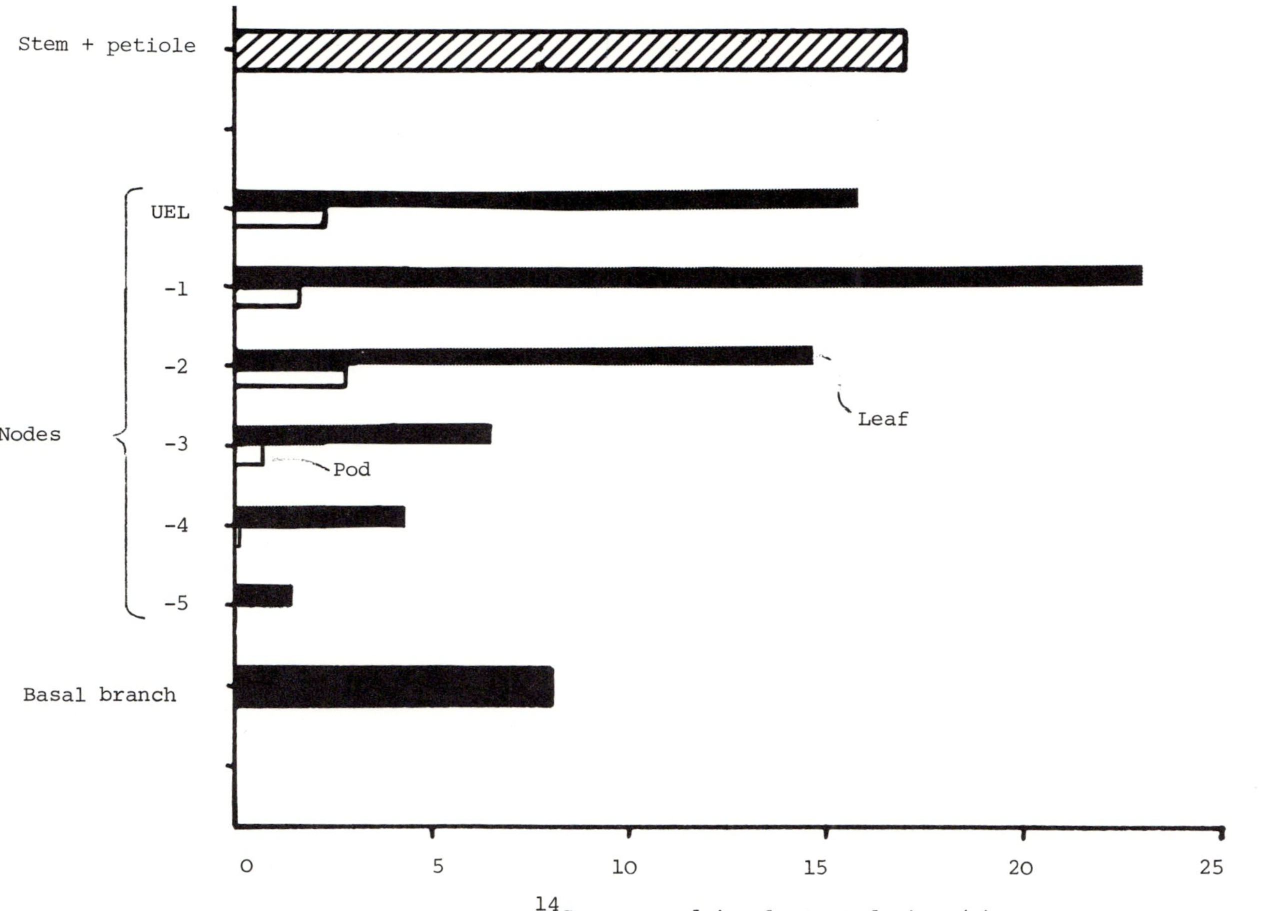

Fig. 3c Percentage contributions of component organs to the photosynthesis of the entire canopy on 2nd August. Topless, irrigated.

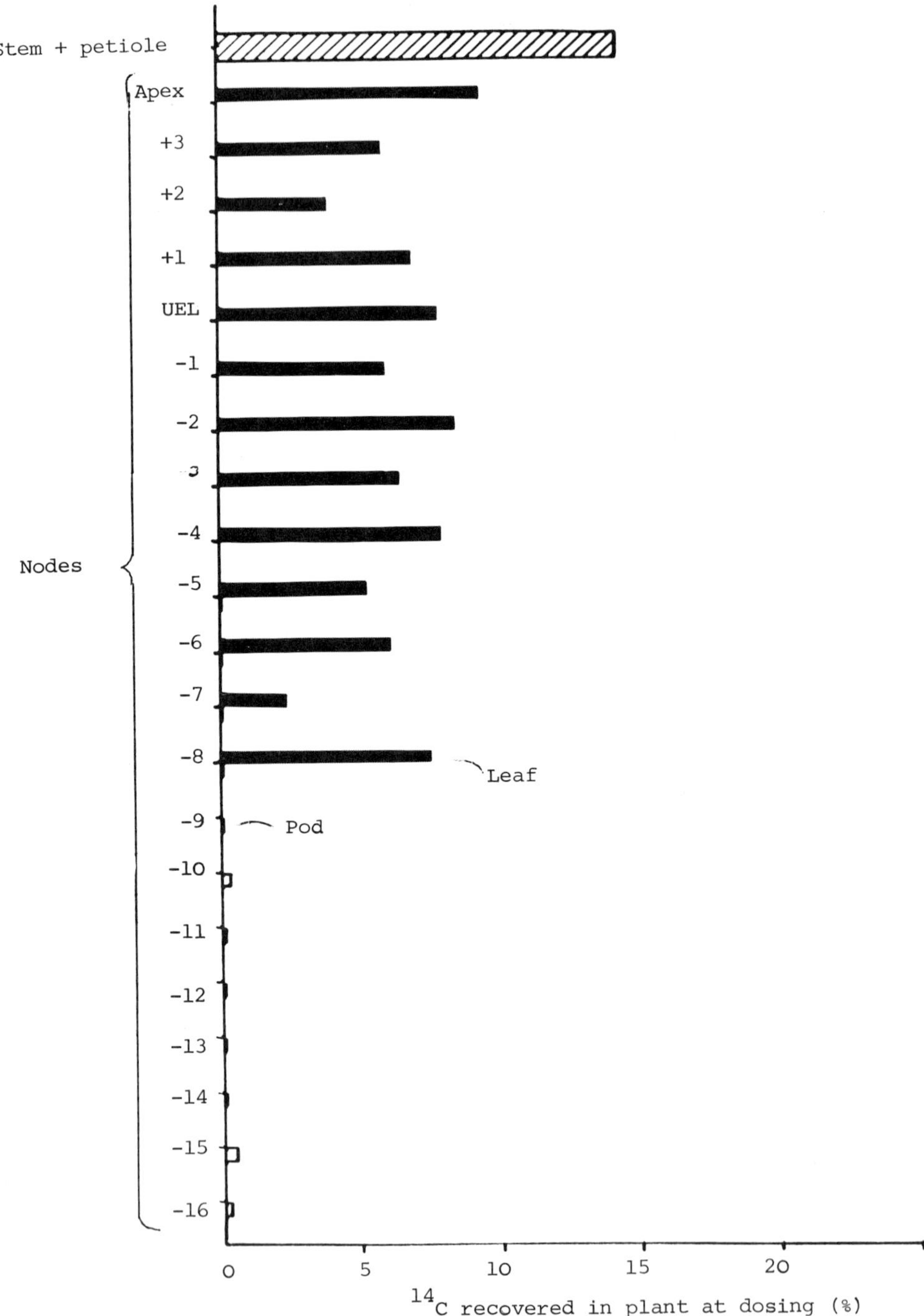

Fig 3d Percentage contributions of component organs to the photosynthesis of the entire canopy on 2nd August.
Maris Bead, irrigated.

genotype, both in terms of branch size and the node from which the branch grew. Branches contributed up to 20% to canopy photosynthesis.

In contrast to Maris Bead the zone of most active photosynthesis was above that of the pods. More leaves contributed to canopy photosynthesis in this genotype and no single leaf was a dominant supplier of assimilate. Also, more leaves were present in the irrigated than in the non-irrigated plots. The pods contributed less in the irrigated plots, partly because they were more shaded by the leaves, but also because the onset of pod growth was later in the irrigated than in the non-irrigated plots (Figures 1 and 2).

The contributions of the pods to canopy photosynthesis at the dosing on 2 August were: non-irrigated: Topless 10.4% Maris Bead 18.2%; irrigated: Topless 7.6%, Maris Bead 1.8%. The specific activity of the pods of Topless was 2 - 3 times those of Maris Bead, indicating that the rate of photosynthesis per unit weight of pod tissue was greater. These differences were related to the size and number of the pods and their exposure to the light.

Translocation of assimilated carbon

As is usual in studies of this kind, the distribution of ^{14}C following dosing of a leaf was taken as a measure of the translocation to various organs. If respiration losses were varying proportions of the carbon imported into different organs, the true patterns of translocation assessed by this method would be distorted.

When the uppermost expanded leaves of non-irrigated Maris Bead plants were dosed before pod filling, 30% of the ^{14}C recovered in the plants five days after labelling was in the stem below the dosed leaves. They retained about 40% of the carbon they had assimilated (Figure 4a). About 15% was translocated to the organs above the dosed leaf and a similar amount

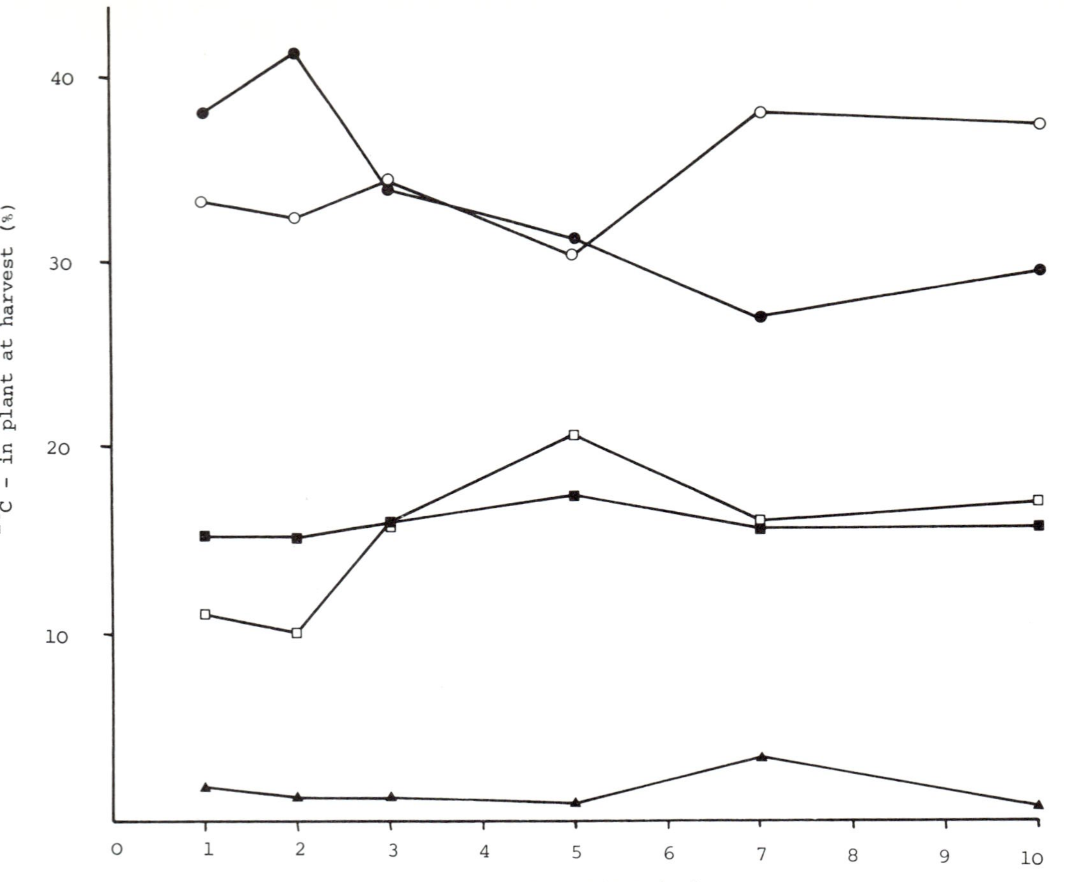

Fig. 4a Change with time after labelling in the distribution of ^{14}C fed to the uppermost expanded leaf (UEL) of Maris Bead. Labelling carried out on 26 June.
Dosed leaf (●), stem below dosed leaf (o), tap root (■), 'apex' - all vegetative organs above UEL (□), leaves below dosed leaf (▲).

to those parts of the root system that were recovered. This pattern of distribution was established within one day of labelling and there was no substantial change during the following 10 days, though there was a gradual loss of ^{14}C from the dosed leaf and a gain by the stem below it and the organs above it.

When the uppermost expanded leaves of non-irrigated Maris Bead plants were dosed during grain filling they retained little ^{14}C and the majority was translocated to the pods within a day. The proportion in the pods gradually increased during the subsequent 10 days (Figure 4b).

On 12 July, at the onset of grain growth, more of the carbon exported from leaves at all levels in the canopy of Maris Bead was translocated upwards in the irrigated than in the non-irrigated plants. When assessed on 26 July, i.e. during grain filling, less of the carbon exported from these leaves was translocated upwards to vegetative organs (Table 2).

TABLE 2

TRANSLOCATION UPWARDS OF ^{14}C FROM DOSED LEAVES OF PLANTS OF MARIS BEAD (% OF TOTAL ^{14}C IN PLANT 5 DAYS AFTER LABELLING)

Position of dosed leaf in canopy	Dosing date and treatment			
	12 July		26 July	
	Not irrigated	Irrigated	Not irrigated	Irrigated
Upper	1.8	22.7	0.3	0.4
Middle	2.5	12.5	1.1	5.6
Lower	2.5	11.6	*	*

* No leaf present

Most of the branches in Topless were infertile, producing no seed. These branches were very variable in size and the node number from which they grew. In the Topless plants studied,

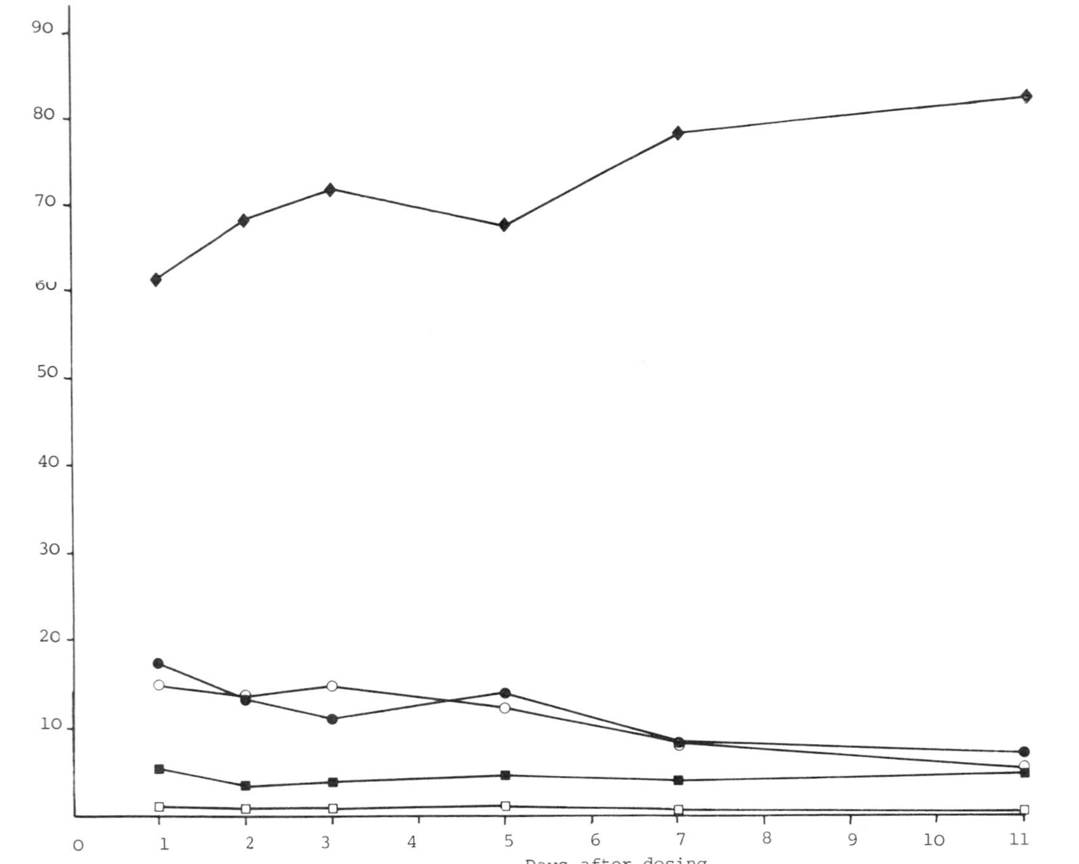

Fig. 4b Change with time after labelling in the distribution of ^{14}C fed to the uppermost expanded leaf (UEL) of Maris Bead. Labelling carried out of 2 August.
Dosed leaf (●), stem below dosed leaf (o), tap root (■), leaves below UEL and 'apex' - all vegetative parts above UEL (□), pods (◆)

branches contributed on average 4.5% to canopy photosynthesis (means of data from 16 July to 2 August) but they were importers of carbon from their mother shoots, receiving an average of 3% (maximum for any plant 14%) of the carbon exported by labelled single leaves on this shoot. Thus, even branches almost as tall as the mother shoot may have been 'parasitic' on the main shoot and did not generally bear ripe seed. Furthermore, it seems certain that they intercepted light which would otherwise have been intercepted by the main shoot.

The distribution and size of pods varied greatly from plant to plant within a plot. Accordingly it is not possible to show how pod size and position in relation to the labelled leaf affected the amount of ^{14}C it received for an 'average' plant. However, from analyses of at least four plants per treatment combination, labelled during the phase of grain growth, the following features were evident:

The patterns of translocation of carbon assimilates were similar for both genotypes and in both experiments when allowance was made for differences in the distribution of pods on the plant.

Pods in the axil of an upper leaf received one third of the total carbon exported from the leaf. Pods in the axil of a lower leaf received three quarters of the carbon exported by the leaf.

Pods on the same side of the plant as a dosed leaf received four fiths of the carbon exported from it to all pods.

Pods nearest to a dosed leaf received most of the carbon exported from it.

Large pods were greater sinks for assimilate than small ones.

CONCLUSIONS

1. Although at maturity Topless produced a similar total amount of dry matter to Maris Bead, it gave a slightly lower yield of dry grain.

2. Frequent irrigation until the onset of pod growth greatly stimulated vegetative growth in both genotypes but increased grain yields only slightly.

3. Both genotypes responded similarly to irrigation in terms of dry matter production.

4. At maturity, Topless plants were shorter than those of Maris Bead, particularly in the irrigated experiment. The yield of vegetative dry matter in Topless depended on the production of tillers and axillary branches. Most branches did not contribute to grain yield and were sinks for assimilate during grain filling.

5. Leaves contributed most to the photosynthesis of the canopy during grain filling. Pods contributed a maximum of 20%.

6. As expected from their position in the canopy, the rate of photosynthesis per unit dry weight of pod was greater in Topless than in Maris Bead.

7. The 'Topless' plant type does not appear to be inherently inferior in productivity to the normal, indeterminate type. It is suggested that in future breeding, selection should be practised for increased production of tillers which develop synchronously with the main shoot and against the production of infertile branches. Genotypes selected to this model may be as high yielding as conventional ones in favourable years while not producing excessive vegetative growth.

REFERENCES

Austin, R.B., Ford, M.A., Edrich, J.A. and Hooper, B.E., 1976. Some effects of leaf posture on photosynthesis and yield in wheat. Annals of Applied Biology 83, 425-446.

Sjödin, J., 1971. Induced morphological variation in *Vicia faba* L. Hereditas 67, 155-180.

ACKNOWLEDGEMENT

We are grateful to D.A. Bond for providing the seed of Topless and to M. Pope for sowing the experiments.

C.L. Hedley *(UK)*

Do you think that you are underestimating the contribution of the pod to photosynthesis, because the pod is producing carbon from within and fixing that carbon rather than atmospheric carbon?

R.B. Austin *(UK)*

There is a possibility. However, I would not have thought it was to a great extent. What I did not mention was that the technique which we use measures the gross photosynthesis of the organs concerned.

C.L. Hedley

But if there is a high concentration of CO_2 within the pod, you then underestimate photosynthesis of the pod, because of the effect of the high CO_2 atmosphere within it.

R.B. Austin

Right. However, I think it is quite likely that, in the case of the field bean, the light intensity inside the pod is very low because of the thickness of the pod. The contribution to photosynthesis by the CO_2 within the pod will be very small indeed.

C.L. Hedley

My colleagues have measured light penetration through pea pods and found it to be high enough to sustain fairly high levels of carbon dioxide fixation.

R.B. Austin

The only useful comparative data that I know of is for wheat where, of course, the pericarp is green and is surrounded by the glumes and the lemmas, etc. The maximum contribution that it seems to be capable of making, even though it has got a C_4 pathway, is about 1% to the photosynthesis of the canopy. However, I would be the first to agree that this is an area on which we do not have adequate information.

D.A. Lawes *(UK)*

Thank you.*

* Additional discussion on this paper appears at the end of this session.

VARIETY X ENVIRONMENT INTERACTION IN *Vicia faba* IN RELATION TO THE GENETIC STRUCTURE OF VARIETIES

J. le Guen and P. Berthelem
Station d'Amelioration des Plantes,
INRA BP29, 35650 Le Rheu, France.

ABSTRACT

Yield stability is an important character to breed into field beans. This stability depends on various parameters among which genetical structure predominates.

Ten varieties, comprising four French winter synthetic, four English winter synthetic, one Franco-English hybrid and one French local population, have been tested in two locations for four years.

Different statistical methods have been used to assess stability among varieties. It seems that stability is more important in winter field beans than it is generally thought to be. There were no important differences between French and English synthetic varieties, except for one of them which was unstable. The results show that the highest yielding varieties are also the most stable.

INTRODUCTION

To be commercially attractive, field beans must give both high and stable yields over quite different environments. In practice, stability, over years for example, is often just as important as yield itself, and it is frequently difficult for the plant breeder to select for stability as this is controlled by several genetic characters.

Several research workers have observed (e.g. for corn or oats) that heterogeneous populations provide the best opportunities to obtain varieties with a small genotype x environment interaction. On the other hand, the environment in which the variety is selected can have an influence on its behaviour in other environments.

So it was interesting to compare some varieties of field beans derived from various geographical origins and of different genetical composition. For this purpose, ten winter field bean varieties were used, comprising four French synthetic varieties, four English synthetic varieties, one Franco-English F_1 hybrid and a French local population very well adapted to its country of origin.

METHODS AND MATERIALS

An experiment was conducted in two localities over four years:

Sites: R = Rennes, D = Dijon
Years: 1975, 1976, 1977, 1978
Varieties:

French synthetic	English synthetic
A = Rovasse	E = Bulldog
B = Survoy	F = Baneer
C = Avrissot	G = Throws MS
D = Soravi	M = Maris Beagle

Franco-English F_1 H = 48B x S45
French landrace L = Cote d'Or

Trials consisted of randomised blocks with five replications in which seed rate was about 25 plants/m^2. Each plot had five rows, 6 m long and 45 cm between rows. At maturity, only the three central rows were harvested. To compare the performance of the varieties grown at several centres in several seasons, three statistical methods were used.

The method described by Finlay and Wilkinson (1963) was used which, for each variety, computes the linear regression of individual yield on the mean yield of all varieties for each site in each season. By this method, an estimate of the effect of the environment is given by the mean yield of all varieties.

The genotypic stability analysis method, as suggested by Tai (1971), was also used. Here, the statistical indexes α and λ are calculated and measure, respectively, the linear response of a variety to environmental effects and the deviation from this linear response. These two indices, with their confidence limits, can be considered as parameters for measuring the stability of a variety.

Finally, an attempt was made to find a way of expressing the specific relationship between a given variety and year or between the variety and locality. Computations were made of the various components of the variance and interactions as partitioned in Table 1. The varieties were ranked on the basis of the values obtained. The difference between the higher and the lower rank in a locality, or for a year, gives an indication of the relative stability of a variety in relation to the particular environmental factor.

RESULTS

Analysis of variance of yields is presented in Table 2. The total sum of squares has been divided into three main effects - genotype, locality and year - and interactions of different orders. Conventional computation of the F-ratios showed all except that for locality to be significant at the 1%

TABLE 1

MODEL OF THE PARTITION OF VARIANCE IN ITS COMPONENTS IN THE SITE X YEAR TRIAL

Model	Component
Yijkl - Y.... = Yi... - Y....	Variety
+ Y.j.. - Y....	Environment
+ Y..k. - Y....	Year
+ (Yi.k.- Yi... - Y..k. + Y....)	Variety x year
+ (Yij.. - Yi... - Y.j.. + Y....)	Variety x environment
+ (Y.jk. - Y.j.. - Y..k. + Y....)	Environment x year
+ (Yijk. - Yi.k. - Y.jk. + Y.k..) - (Yij.. - Yi... - Y.j.. + Y....)	Variety x environment x year
+ (Y.jkl - Y.jk.)	Repetition
+ Yijkl - Yijk. - Y.jkl - Y.jk.	Residual

Yijkl = yield of the ith variety in the jth location of the kth year in the lth replication

TABLE 2

ANALYSIS OF VARIANCE OF YIELDS (Qx/ha AT 15% MOISTURE CONTENT) OF TEN VARIETIES OF WINTER BEAN IN 2 LOCATIONS FOR 4 YEARS

Source of variation	d.f.	M.S.
. Total		
. Genotype	9 (G-1)	862.8**
Genotype x year	27 (G-1)(A-1)	1 946.7**
Genotype x site	9 (G-1)(L-1)	985.1**
Genotype x year x site	27 (G-1)(L-1)(A-1)	2 854.3**
. Environment	3 (A-1)(L-1)	21 212.9**
Site	1 L-1	73.4^{NS}
Year	3 A-1	14 067.8**
. Repetition in environment	32 E (R-1)	674.5**
. Error	288 E (R-1)(G-1)	43.0

** Significant at 1% probability level

level. It is interesting to note the very high value of the mean square for years, which is sixteen times bigger than the genotypic one.

Table 3 gives the values of the mean yields (quintals/ha at 15% moisture content) of the ten varieties studied for four years at two sites.

The lower half of Table 3 gives mean values for years averaged over sites. This may not be strictly correct because the variety x site interaction was significant but it was only one half the value of that for variety x year and the main effect of locality was not significant. (It is shown later that this modification had little effect on the calculated coefficients).

From the values in Table 3, estimates were made of the coefficients of regressions between the individual yield values and the mean yields in each location for each year (Finlay and Wilkinson, 1963). Values for correlation coefficients (R), the intercept 'a' and slope 'b' are given in Table 4. The upper half of the Table corresponds to the upper half of Table 3 in which sites and years were separate, and the lower half of the Table corresponds to the lower half of Table 3 giving averages over sites.

In this type of analysis a regression coefficient of the order of 1.0 indicates an average level of stability over all environments. Clearly the value (Table 4) for the majority of the varieties was practically equal to 1.0 except for varieties F and L.

The results are shown graphically in Figure 1 where the abscissa is the environment mean yield and the ordinate is the yield of individual varieties. The dotted line represents the population mean which has a coefficient of 1. Only four varieties are shown - H, A, F and L - characterising four different responses.

TABLE 3

MEAN YIELDS (Qx/ha AT 15% MOISTURE CONTENT) OF 10 WINTER FIELD BEAN VARIETIES IN 4 CONSECUTIVE YEARS AT 2 DIFFERENT SITES

Site	Rennes				Dijon				
Year	1975	1976	1977	1978	1975	1976	1977	1978	Mean
Variety									
A	37.2	42.3	33.7	59.2	62.6	23.0	27.2	55.2	42.5
B	33.2	41.5	31.9	55.9	62.4	28.9	26.0	55.7	42.0
C	33.9	42.9	37.6	55.7	66.8	25.5	27.6	55.0	43.1
D	37.6	41.2	34.1	56.2	55.6	25.0	26.0	53.4	41.2
E	39.8	46.0	30.3	59.9	59.7	26.2	21.9	50.6	41.8
F	36.5	44.9	26.7	62.8	59.5	27.0	16.0	53.1	40.8
G	37.8	40.8	29.9	57.3	56.5	23.6	22.8	49.3	39.7
H	38.0	53.3	40.8	60.3	64.5	34.8	31.3	57.0	47.5
L	28.1	35.7	16.8	23.0	42.4	24.7	17.2	54.1	29.1
M	37.2	43.6	29.7	53.7	55.0	23.2	20.9	54.4	39.7
Mean	35.9	43.2	31.2	54.4	58.5	26.2	23.7	52.9	40.7

MEAN YIELDS (Qx/ha AT 15% MOISTURE CONTENT) OF 10 WINTER FIELD BEAN VARIETIES IN 4 YEARS AVERAGED OVER SITES

Year	1975	1976	1977	1978
Variety				
A	49.9	32.6	30.4	57.2
B	47.8	35.2	29.0	55.8
C	50.4	34.2	32.6	55.3
D	46.6	33.1	30.1	54.8
E	49.7	36.1	26.1	55.2
F	48.0	35.9	21.3	58.0
G	47.1	32.2	26.3	53.3
H	51.2	44.1	36.0	58.6
L	35.2	30.2	17.0	34.1
M	46.1	33.4	25.3	54.0
Mean	47.2	34.7	27.4	53.6

TABLE 4

COEFFICIENTS OF THE REGRESSION BETWEEN INDIVIDUAL YIELDS (Qx/ha AT 15% MOISTURE CONTENT) OF 10 WINTER BEAN VARIETIES (A TO M) AND THE MEAN YIELD OF THE POPULATIONS GROWN FOR 4 YEARS IN 2 LOCATIONS

	Variety									
	A	B	C	D	E	F	G	H	L	M
R	0.9751	0.9747	0.9441	0.9750	0.9684	0.9581	0.9730	0.9592	0.5384	0.9816
Intercept a	-2.0676	-0.3329	-0.2555	3.5210	-2.0142	-9.5061	-1.3472	9.7925	5.0169	-2.5410
Slope b	1.0949	1.0373	1.0646	0.9250	1.0752	1.2348	1.0085	0.9253	0.5916	1.0369

COEFFICIENTS OF THE REGRESSION BETWEEN INDIVIDUAL YIELDS (Qx/ha AT 15% MOISTURE CONTENT) OF 10 WINTER BEAN VARIETIES (A TO M) AND THE MEAN YIELD OF THE POPULATION IN 4 YEARS AVERAGED OVER SITES

	Variety									
	A	B	C	D	E	F	G	H	L	M
R	0.9662	0.9960	0.9587	0.9752	0.9939	0.9802	0.9956	0.9811	0.7519	0.9988
Intercept a	-1.6280	0.3845	4.6474	1.7808	-3.3123	-12.9426	-3.3479	14.5188	4.1726	-4.3227
Slope b	1.0841	1.0202	0.9442	0.9664	1.1074	1.3191	1.0576	0.8091	0.6124	1.0806

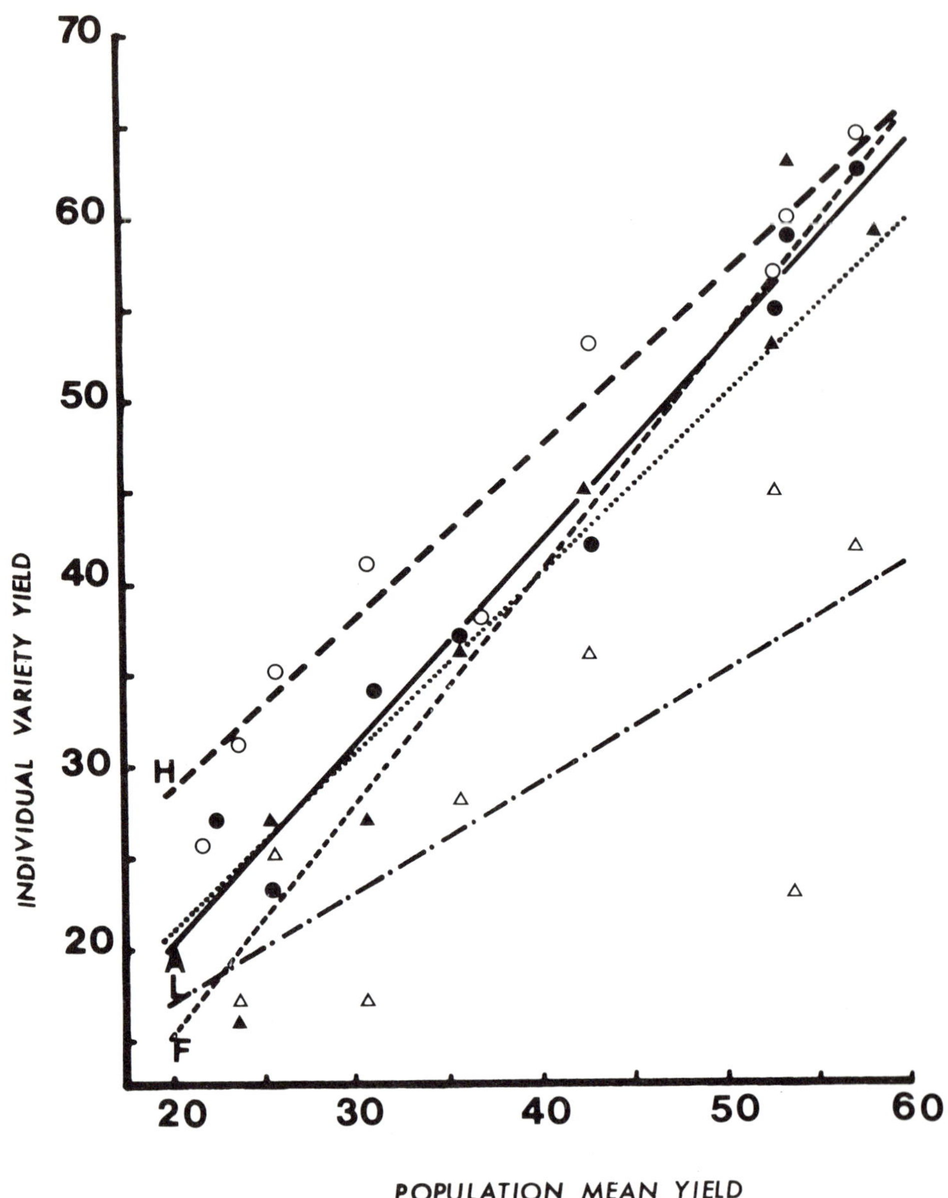

Fig. 1: Relationship between environment means and individual variety means for yield.

Variety L produced below average yields in all seasons at all sites. Furthermore, this variety showed relatively little change in yield despite changes in the environment. Variety L can be considered as being poorly adapted to all environments.

Conversely, the F_1 hybrid variety H always produced an above-average yield, indicating that it possesses a general adaptability and average stability as shown by its correlation coefficient of about 1.0.

Variety F gave a correlation coefficient statistically greater than 1.0. It was of below-average stability and was very sensitive to change in environment, performing worst in the poorest environment and best in the most favourable. So, this variety was unstable and especially adapted to high-yielding environments.

The parameters for variety A compare closely to those for the mean population, as do those of the other synthetic varieties and, thus, they may be considered as of average stability over different environments.

To confirm these results, the genotypic stability analysis method of Tai (1971) was used, and consists principally of partitioning the genotype x environment interaction effect into two components. These are the linear response to environmental effects, measured by the α statistic, and the deviation from the linear response, measured by the λ statistic.

It has been shown by Tai that α and λ can be easily computed from the environmental effect $\hat{l}j = \bar{x}.j. - \bar{x}....$, from the genotype x environment interaction effect $(\widehat{gl})_{ij} = (\bar{x}ij. - \bar{x}i.. - \bar{x}j.. - \bar{x}...)$ and from the value of the mean squares derived from the analysis of variance.

The prediction interval for $\alpha = 0$ and the upper limits of the confidence interval for $\lambda \geqslant 1$ are determined respectively from the 't law' and from the 'F law'. Values for varieties within these limits can be postulated to represent a high level of stability. Table 5 gives the values of the different parameters computed, with an indication of significant effects, and the data are presented graphically in Figure 2 showing confidence intervals for the α statistic at probability levels of 0.9 and 0.95. The value of λ for variety L is outside the scale used in the diagram, confirming its high instability shown previously. Also, the value for variety F was greater than the highest limit for the confidence interval, and that for variety M was smaller than that for the lower limit for this parameter, indicating instability in each case. The other varieties are, as previously seen, all within error and can be considered as stable over all the environments tested.

In order to see which environmental factor (year or site) most affected yield, the interaction for each variety with the relevant component was calculated. Consequently, four values (1 for each year) per variety were obtained for variety x year interaction ($k = 1, 2, 3, 4$) and two values per variety in the case of variety x site interaction ($j = 1, 2$), (Tables 6 and 7). Varieties have been ranked according to the value of their individual interaction. Differences were then calculated between the highest and lowest rank values for each variety, (Table 7). Classification of the varieties on this basis is interpreted as representing the relative response of a given variety to a given factor. Ranking of varieties on the basis of variety x year interaction differed from that based on variety x location interaction, with the exception of varieties L and M. Figure 3 shows the relationship between ranking on the basis of the interaction for variety x year (abscissa) and on the basis of the interaction for variety x site (ordinate).

By constructing a line from the middle of the X and Y axes, the plane (X - Y) is divided into quarters. The position given by the ordinates of each variety within the plane (X - Y) may be used to indicate the relative importance of each environmental factor. In the upper left quarter there is little or no

TABLE 5

P VALUES FOR LINEAR RESPONSES (α_i) AND DEVIATIONS (λ_i) FOR 10 VARIETIES IN RELATION TO ENVIRONMENTAL EFFECTS

Site		Rennes				Dijon			
Year		1975	1976	1977	1978	1975	1976	1977	1978
$\hat{l}$		-4.84	2.46	-9.54	13.66	17.76	-14.54	-17.04	12.16
$\widehat{(gl)}$	A	-0.50	-2.70	0.70	3.00	2.30	-5.00	1.70	0.50
	B	-4.00	-3.00	-0.60	0.20	2.60	1.40	1.00	1.50
	C	-4.40	-2.70	4.00	-1.10	5.90	-3.10	1.50	-0.30
	D	1.20	-2.50	2.40	1.30	-3.40	-1.70	1.80	0.00
	E	2.80	1.70	-2.00	4.40	0.10	1.10	-2.90	-3.90
	F	0.50	1.60	-4.60	8.30	0.90	0.70	-7.80	0.10
	G	2.90	-1.40	-0.30	3.90	-1.00	-1.60	0.10	-2.60
	H	-4.70	3.30	2.80	-0.90	-0.80	1.80	0.80	-2.70
	L	3.80	4.10	-2.80	-19.80	-4.50	10.10	5.10	3.80
	M	2.30	1.40	-0.50	0.30	-2.50	-2.00	-1.80	2.50

	Mean yield Qx/ha	$\hat{\alpha}$	$\hat{\lambda}$
A	42.5	0.042	0.838
B	42.0	0.017	0.675
C	43.1	0.029	1.629
D	41.2	-0.029	0.561
E	41.8	0.022	0.995
F	40.8	0.104	2.288*
G	39.7	0.004	0.664
H	47.5	-0.033	0.928
L	29.1	-0.181**	9.225*
M	39.7	0.016	0.407*

* = λ greater than Fa for $n_1 = n-2$ and $n_2 = n(m-1)(p-1)$ degrees of freedom at the $\alpha = 0.05$ level

** = α significantly different from 0 at the 5% level

$\hat{l}$ = environmental effect

$\widehat{(gl)}$ = genotype x environment effect

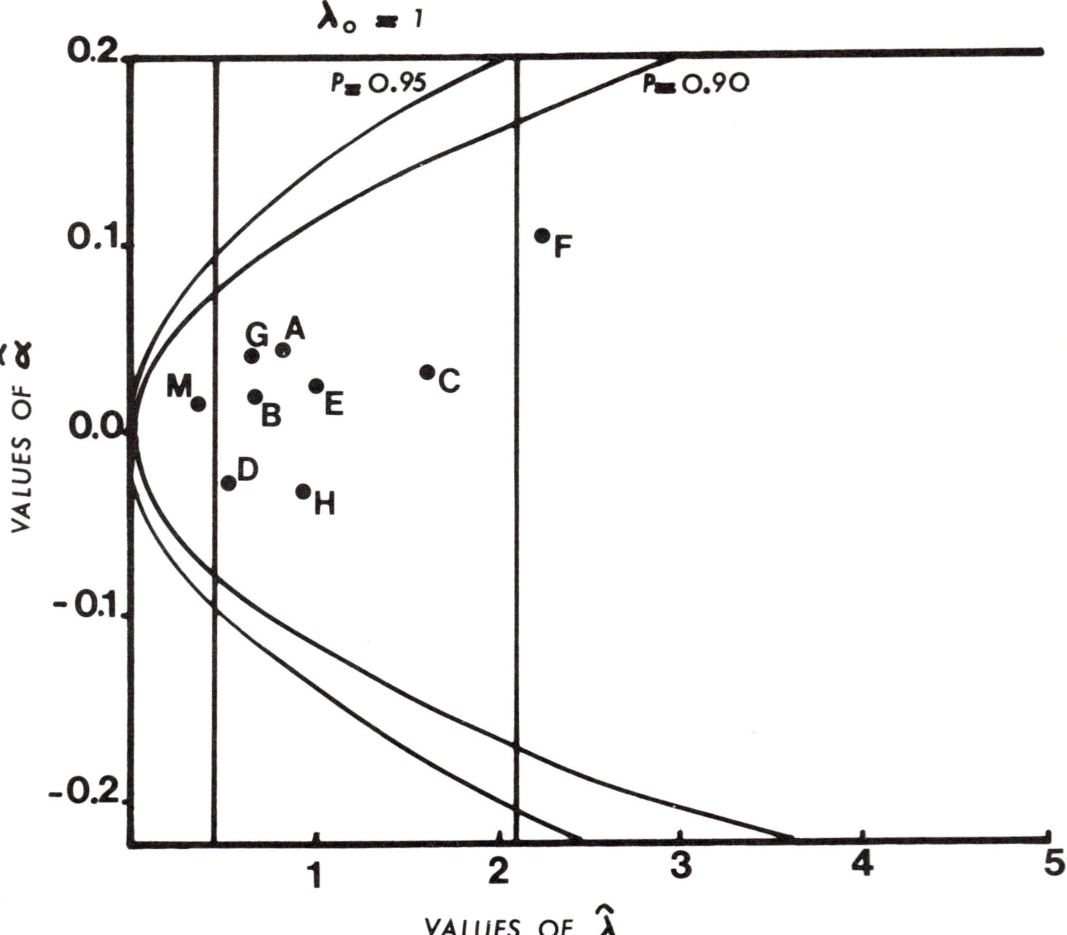

Fig. 2: Relationship between α and λ for 9 of the 10 varieties used.

TABLE 6

CLASSIFICATION OF 10 VARIETIES ON THE BASIS OF YEAR

Interaction: variety x year				
Variety	Year	Value of Yi.k.-Yi...-Y.k.+ Y...	Rank of the variety	Mean of the variety
A	1	+ 0.88	26	49.88
	2	- 3.87	3	32.64
	3	+ 1.22	30	30.44
	4	+ 1.76	35	57.20
B	1	- 0.60	14	47.81
	2	- 0.71	12	35.21
	3	+ 0.34	19	28.96
	4	+ 0.97	28	55.82
C	1	+ 0.77	25	50.35
	2	- 2.93	5	34.16
	3	+ 2.81	38	32.61
	4	- 0.68	13	55.34
D	1	- 0.99	11	46.63
	2	- 2.01	8	33.12
	3	+ 2.20	36	30.06
	4	+ 0.75	24	54.81
E	1	+ 1.46	33	49.73
	2	+ 0.35	20	36.13
	3	- 2.35	6	26.14
	4	+ 0.52	22	55.23
F	1	+ 0.74	23	48.00
	2	+ 1.30	31	35.90
	3	- 6.15	2	21.33
	4	+ 4.25	39	57.95
G	1	+ 0.93	27	47.13
	2	- 1.49	10	32.22
	3	- 0.08	18	26.34
	4	+ 0.64	21	53.28
H	1	- 3.10	4	51.22
	2	+ 2.23	37	44.06
	3	+ 1.48	34	36.02
	4	- 2.12	7	58.64
L	1	- 0.38	15	35.21
	2	+ 7.14	40	30.24
	3	+ 1.16	29	16.97
	4	- 7.95	1	34.08
M	1	- 0.09	17	46.07
	2	- 0.25	16	33.42
	3	- 1.90	9	25.29
	4	+ 1.44	32	54.04

Yijkl = yield of the ith variety in the jth site of the kth year in the lth replication.

TABLE 7

CLASSIFICATION OF 10 VARIETIES ON THE BASIS OF SITE

Interaction: variety x location				
Variety	Location	Value of Yij..-Yi...-Y.j..+ Y....	Rank of the variety	Mean of the variety
A	1	+ 0.12	12	43.09
	2	- 0.12	11	42.00
B	1	- 1.74	3	40.64
	2	+ 1.73	18	43.26
C	1	- 1.02	6	42.53
	2	+ 1.00	15	43.70
D	1	+ 0.71	13	42.30
	2	- 0.72	8	40.02
E	1	+ 1.76	19	44.00
	2	- 1.77	2	39.62
F	1	+ 1.47	17	42.70
	2	- 1.47	4	38.91
G	1	+ 1.28	16	41.45
	2	- 1.28	5	38.04
H	1	- 0.19	10	48.10
	2	- 0.56	9	46.88
L	1	- 3.66	1	25.90
	2	+ 3.65	20	32.36
M	1	+ 0.90	14	41.03
	2	- 0.89	7	38.39

CLASSIFICATION OF VARIETIES IN RELATION WITH THE DIFFERENCE BETWEEN THE MAXIMUM RANK AND MINIMUM RANK IN A GIVEN LOCATION AND YEAR

	V x A		V x L	
	Δ |max - min|	Rank	Δ |max - min|	Rank
A	32	4	1	9
B	16	10	15	3
C	33	3	9	6
D	28	6	5	8
E	27	7	17	2
F	37	2	13	4
G	17	9	11	5
H	30	5	1	9
L	39	1	19	1
M	23	8	7	7

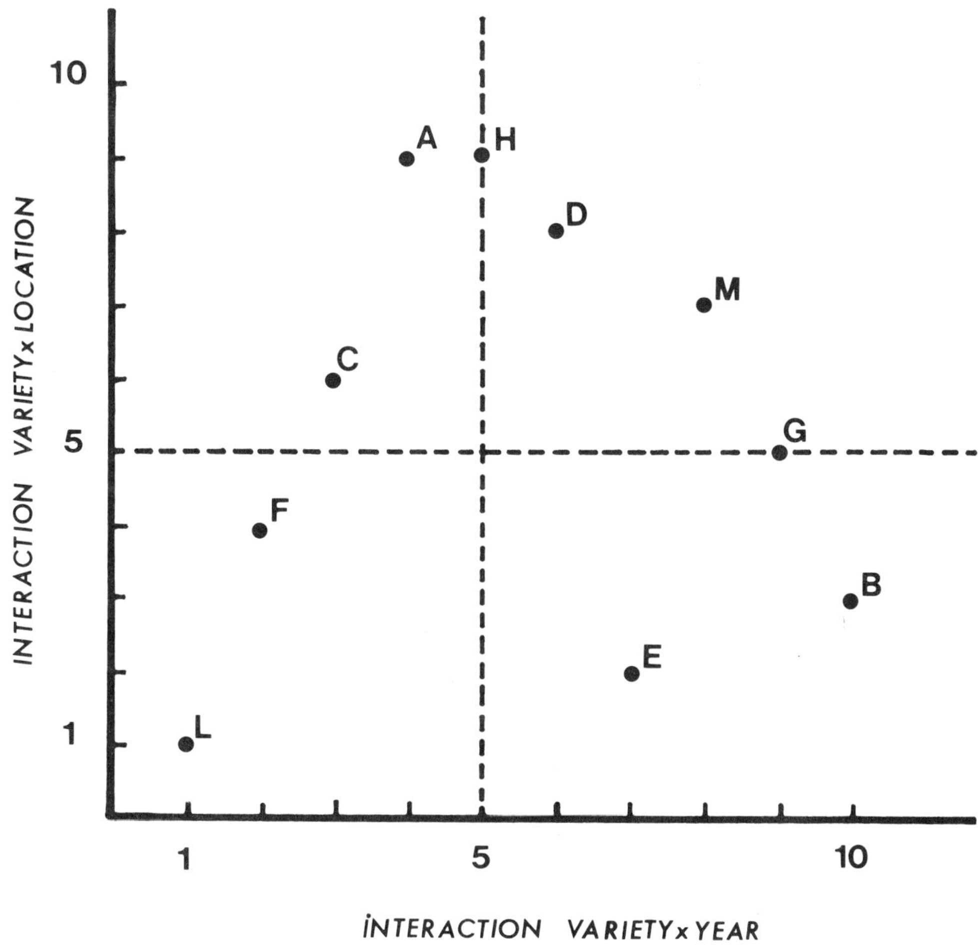

Fig. 3: Relationship between ranking of varieties on the basis of variety x year and variety x site interactions.

variety x locality interaction but a high variety x year one. The varieties A, H and C fall into this category. In the upper right quarter, there is little interaction, either variety x year or variety x location, and this applies to the varieties D and M. In the lower right quarter, there is a variety x location but no variety x year interaction, and here the varieties G, E and B are represented. Finally, in the lower left quarter there is both a variety x year and a variety x locality interaction, which included the varieties F and L.

DISCUSSION

The different methods presented in this paper have given an idea of the relative stability of various varieties of field bean in several environments.

The weather differed markedly between the years included in the study; 1975 and 1976 had especially dry periods during spring and summer and relatively high temperatures in summer. Conversely, 1977 and 1978 were very wet years with temperatures generally lower than the annual mean, both during spring and summer. Rennes and Dijon also represented somewhat different climates, Dijon having the higher summer temperatures, lower winter ones, less rain and practically no *Botrytis fabae*. So clearly, the range of environments was sufficiently broad to provide for a substantial test of the varieties. Within these limits, it seems that the varieties tested, with the exception of two, were relatively stable and certainly more stable than is generally thought.

Although the number of varieties was relatively small, and only one F_1 hybrid and one population were included, it is reasonable to conclude that the performance of these two types of varieties differed. The stability of the F_1 hybrid variety and its good performances in all the environments tested is inconsistent with the opinion that heterogenous populations are more stable than homogenous ones. Futhermore, it seems that the level of yield is correlated with the stability of the variety, the highest yielding varieties being also the most stable.

It is difficult, clearly, to identify differences between geographical groups of synthetic varieties. However, it seems that French synthetic varieties were more often affected by year (e.g. varieties A, H, C) than by location (variety B), and, conversely, English synthetic varieties were, perhaps, more affected by location (varieties F, E, G) than by year.

REFERENCES

Finlay, K.W. and Wilkinson, G.N., 1963. The analysis of adaption in a plant breeding programme. Aust. J. Agric. Res., 14, 742-754.

Tai, G.C.C., 1971. Genotypic analysis and its application to potato regional trials. Crop. Sci., 11, 184-190.

DISCUSSION

D.A. Bond *(UK)*

Is there any common morphological feature of the two unstable varieties Baneer and Côte d'Or that you could relate to their instability? Baneer, for example, is more early maturing than the other three English synthetic varieties.

J. le Guen *(France)*

No, we have no data on morphological aspects, our data relates only to yield.

D.A. Bond

The main cause of low yields in the poor environments was the dry years of 1975/76, so these so-called unstable varieties are low-yielding in dry years but they respond to wet years.

J. le Guen

As you can see from my Table 3, the mean for 1976 was not the lowest, in fact, in 1977, the general mean was lower than in 1976. As you can see, 'F' was not too bad in 1976.

E. von Kittlitz *(FRG)*

If I have understood you correctly, you are saying that the most stable variety is the variety with the highest yield. Is that correct? Is that irrespective of whether it is a synthetic or a hybrid variety?

J. le Guen

Yes, the F_1 variety is one of the most stable ones, and this variety is also the highest yielding.

E. von Kittlitz

I think much more work is required on this matter because it is very important to know whether hybrids are more stable than other varieties.

THE EFFECT OF INITIAL SEED MOISTURE CONTENT ON GERMINATION, STAND AND YIELD OF FABA BEANS *(Vicia faba)* AND PEAS *(Pisum sativum)* IN WESTERN CANADA

G. G. Rowland
Crop Development Centre, University of Saskatchewan,
Saskatoon S7N OWO, Canada

ABSTRACT

Four faba bean (Vicia faba *L) and four pea* (Pisum sativum *L) cultivars of varying seed moisture levels were examined for germination, seedling dry weight, seed leakage, field emergence and field performance. Little relationship was found between the laboratory results and the field performance. It was suggested, however, that plant breeders routinely perform germination tests at 5°C and conductivity tests for several years prior to the release of a cultivar. This would help to prevent the release of cultivars with extremely poor germination characteristics.*

INTRODUCTION

Experiments on germination failure in faba beans (*Vicia faba* L.) and peas (*Pisum sativum* L.) were begun at the Crop Development Centre after the spring of 1973. In that year stands of the faba bean cultivars Maris Bead and Ackerperle and the pea cultivar Triumph were very poor. Rowland and Gusta (1977) showed that there were significant differences among faba bean and pea cultivars in the total amount of water imbibed and the amount of leakage of solutes from the seed after 24 h of soaking. In faba beans the cultivars Ackerperle and Maris Bead had the greatest seed leakage and the greatest transverse cracking of the cotyledons (TVC). These were also the cultivars that performed the poorest in the germination tests. The smooth-seeded pea cultivars Triumph and Century had a significantly greater amount of seed leakage than the wrinkle-seeded lines P1244095 and P1280064. However, in peas seed leakage was not as good a predictor of germination failure as in faba beans. Century, which had the highest germination percentage, and Triumph, which had the lowest, both exuded large amounts of solutes.

Since the above series of experiments were not carried out in conjunction with field trials it was not known how the results might relate to events in the field. In addition, the measurements of seed leakage were only performed at one temperature (20°C It was not know what differences, if any, might occur at lower temperatures. Hobbs and Obendorf (1972) found that the loss of cellular constituents in soya beans (*Glycine max*) was most severe in low-moisture seed at low temperatures. Therefore, it was decided to carry out a series of experiments to examine the effec of varying water inbibitional temperatures and seed moisture levels on germination and seed leakage. In addition, the same seed and seed moisture levels were used for seeding date experiments sown at Saskatoon.

MATERIALS AND METHODS

The same four cultivars of faba beans and four cultivars of

peas used in previous experiments (Rowland and Gusta, 1977) were used here. The faba beans, Ackerperle, Diana, Erfordia and Maris Bead, are grown commercially in Western Canada. Of the pea cultivars, Century is a medium-seeded and Triumph a large-seeded round pea licensed for production in Canada. PI244095 is a large-seeded and PI280064 a small-seeded wrinkled pea orginally obtained from the United States Department of Agriculture. All eight were multiplied in 1977 and only seed from this multiplication was used in the 1978 experiments. In 1978 the four pea lines were multiplied for the 1979 experiments, but an error in the faba bean multiplication prevented that seed being used. Seed of the four faba bean cultivars was therefore bulked from the 1978 field experiments and used for the 1979 experiments.

SEED MOISTURE

For the 1978 experiments an attempt was made to adjust the initial seed moisture to 7, 14 and 21% respectively, by adding the appropriate amount of distilled water to seeds in plastic bags. The seed was then allowed to take up this moisture at room temperature for at least 24 h before being used in any of the experiments. This resulted in average seed moistures of 8, 13.5 and 19% respectively. In 1979 no water was added to seed for the low temperature category. The average seed moisture was 5, 12.5 and 17.5 respectively.

As the seed moisture content varied slightly from cultivar to cultivar and year to year, they are reported in the text as being either 7, 14 or 21%.

LABORATORY EXPERIMENTS

Trays containing a sand-sawdust mixture were placed in incubators at 5, 10 or 15°C, for at least 24 h. Four replicates of 35 seeds/tray for each moisture level and cultivar combination were then placed in each incubator and allowed to imbibe water for 48 h. The temperature of the incubators was then raised to

$20^{o}C$. After 7 days the trays were removed and the number of normal seedlings counted. The dry weights of the seedlings were determined after the cotyledons had been removed.

Distilled-deionised water was placed in 125 ml flasks in incubators at 5, 10 or $15^{o}C$ for 48 h. Three replicates of 10 seeds for each cultivar, each moisture level and each temperature were then placed in these flasks for 24 h. The flasks were then removed and conductivity values of the seed leachate obtained with a conductivity meter.

FIELD EXPERIMENTS

The field experiments were planted in a split-split-plot design with seeding dates as the main plots, seed moisture levels as the sub-plots and cultivars as the sub-sub-plots. The seeding dates were May 1, May 15 and June 1 in 1978 and May 9, May 24 and June 8 in 1979.

Thermocouples were placed in the soil, 8 cm deep, at four locations within the field to obtain data on soil temperatures. Readings were taken at 08.00 (night temperature) and 14.00 (day temperature) in varying sequences of days from the first planting date until the last.

RESULTS

Laboratory experiments

In 1978 changes in seed moisture level affected faba bean germination at 5 and $10^{o}C$ (Figure 1). At the two lower temperatures germination was best with a seed moisture level of 14%. However, the cultivars behaved differently, some showing increased germination and others not. The 1979 results were quite different with generally fewer normal seedlings at the 14% seed moisture level than at the other two moisture levels at all temperatures (Figure 2). The cultivars were much more consistent in their response as a significant cultivar x moisture level interaction was found only at $10^{o}C$. The total number of normal faba bean

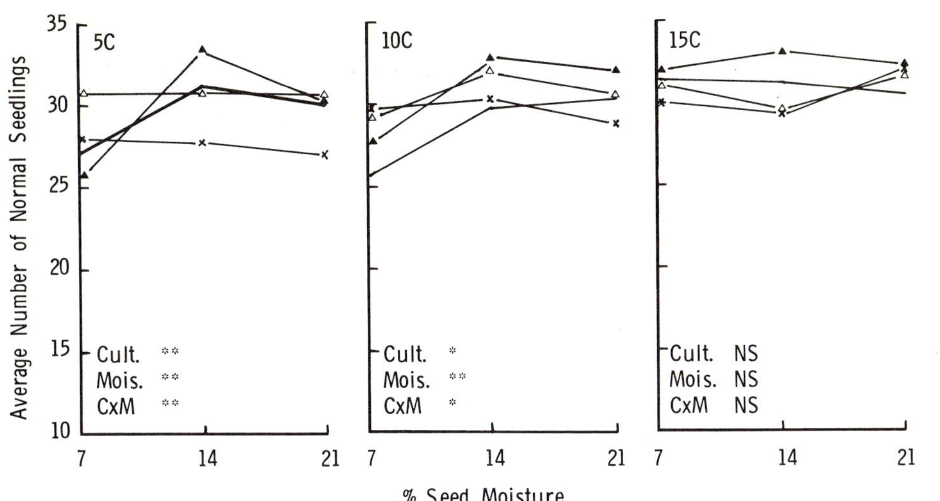

Fig. 1 The average number of faba bean seedlings after 7 days for the cultivars Ackerperle (————), Diana (x————x), Efordia (Δ————Δ) and Maris Bead (▲————▲) for a 48 h imbibitional temperature of 5, 10 or 15°C at varying initial seed moisture levels in a) 1978, and b) 1979. ** and * significant at the 1 and 5% level of probability respectively. NS: non-significant.

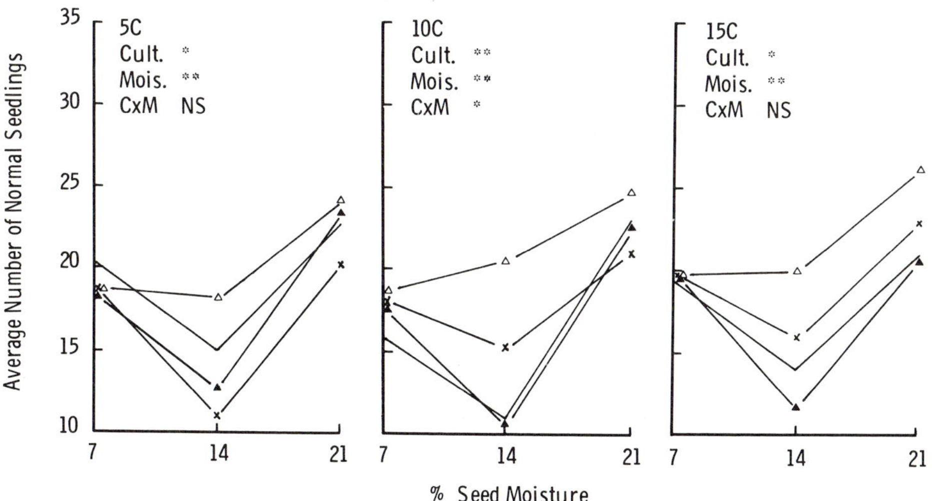

Fig. 2 The average number of normal faba bean seedlings in 1979.

seedlings was much higher in 1978 than 1979. Ackerperle and Maris Bead had much lower numbers of normal seedlings at the 14% moisture level at all temperatures in 1979. This is in contrast to the 1978 tests where Maris Bead had a greater number of normal seedlings.

Seed moisture levels had no significant effect on the average number of normal pea seedlings in 1978 (Figure 3). With the exception of the 15°C temperature, Triumph had the fewest normal seedlings. In 1979 significant differences in germination were found among seed moisture levels at 5 and 10°C and significant cultivar x moisture level interactions at both these temperatures (Figure 4). The smooth-seeded cultivars Triumph and Century were more erratic in their response than the wrinkle-seeded types. Triumph had fewer normal seedlings at each temperature at the 14% moisture level than any other variety. As was the case with faba beans the 1979 germination results were much lower and more variable than in 1978.

The faba bean cultivars differed significantly in seedling dry weights in 1978 with Maris Bead having the lowest average weight at all temperatures and moisture levels (Figure 5). A significant drop in seedling dry weight occurred at the 14% moisture level at 10 and 15°C, but a significant interaction at 10°C showed that all cultivars were not the same in this response. The average seedling dry weight was also greater at 10°C than at either 5 or 15°C. In 1979 only moisture levels at 10°C and cultivars at 5°C had a significant effect on seedling dry weight (Figure 6). The average seedling dry weights were also much lower than in 1978 and were much the same at all temperatures.

The two large-seeded pea cultivars, P1244095 and Triumph generally had the greater seedling dry weights in 1978 (Figure 7). A significant drop in seedling dry weight occurred at the 14% moisture level at 15°C. Century and P1244095 were the most variable cultivars in this character. Significant cultivar differences were present at all temperatures in 1979 with P1244095 once again having the highest average seedling weight (Figure 8).

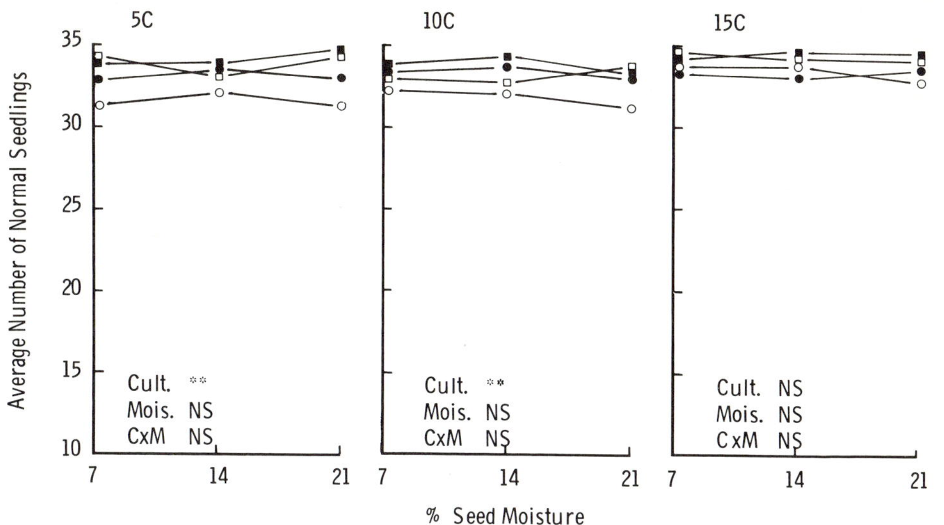

Fig. 3 The average number of normal pea seedlings after 7 days for the cultivars Triumph (o——o), Century (●——●), PI280064 (□——□), and PI244095 (■——■) for a 48 h imbibitional temperature of 5, 10 or 15°C at varying initial seed moisture levels in a) 1978, and b) 1979. ** significant at the 1% level of probability, NS: non-significant.

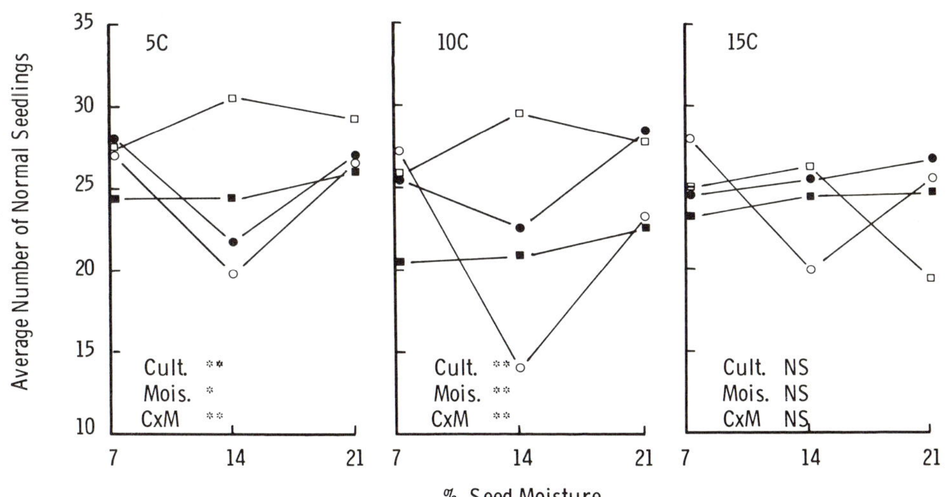

Fig. 4 The average number of normal pea seedlings in 1979. (see caption Figure 3).

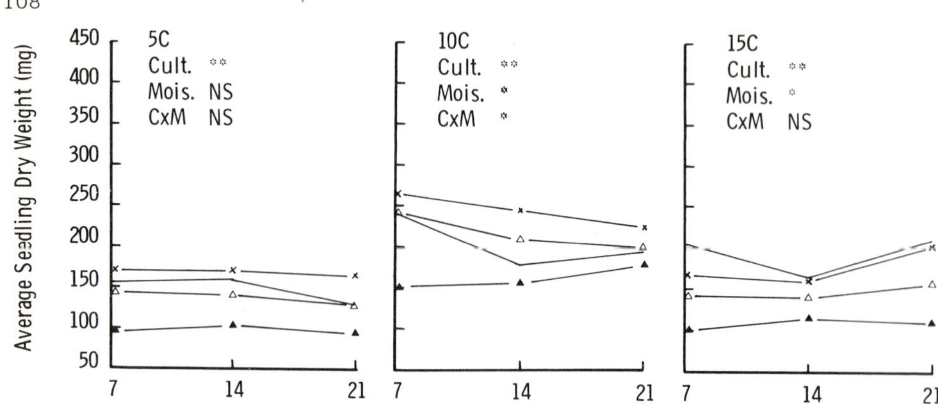

Fig. 5 The average faba bean seedling dry weights after 7 days for the cultivars Ackerperle (————), Diana (x———x), Erfordia (Δ———Δ) and Maris Bead (▲———▲) for a 48 h imbibitional temperature of 5, 10 or 15°C at varying seed moisture levels in 1978. ** and * significant at the 1 and 5% level of probability, respectively. NS: non-significant.

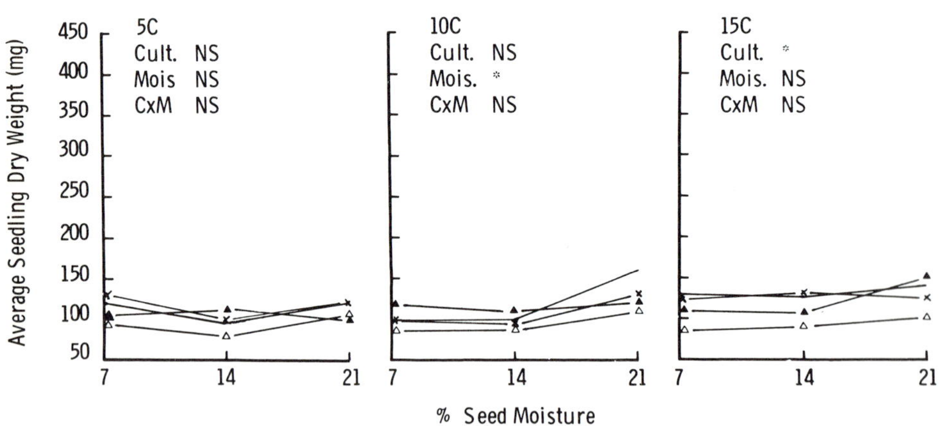

Fig. 6 The average faba bean seedling dry weights in 1979. (See caption Figure 5).

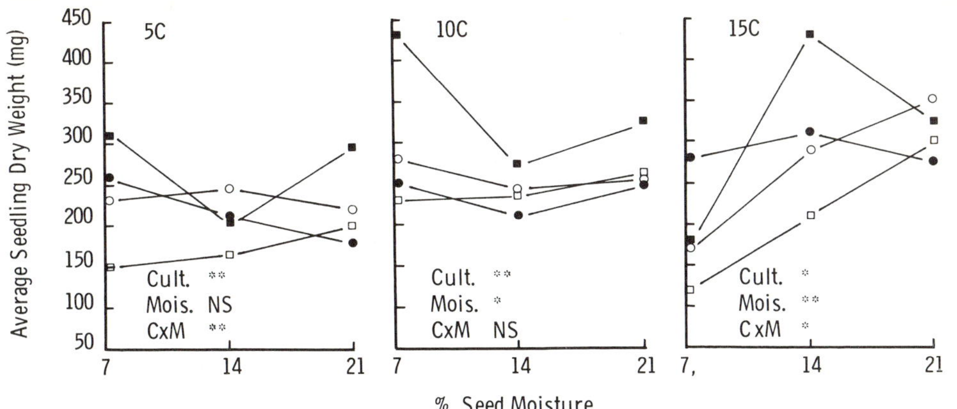

Fig. 7 The average pea seedling dry weight after 7 days for the cultivars Triumph (o——o), Century (●——●), PI280064 (□——□), PI244095 (■——■) for a 48 h imbibitional temperature of 5, 10 or 15°C at varying seed moisture levels in a) 1978 and b) 1979. ** and * significant at the 1 and 5% level of probability, respectively. NS: non-significant.

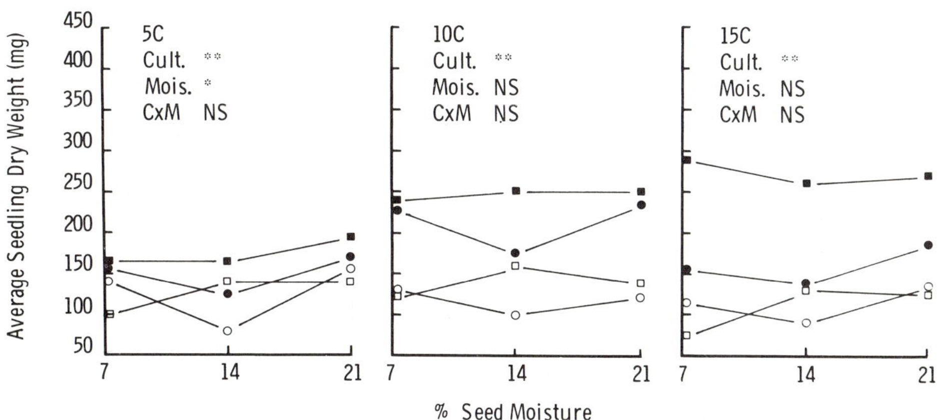

Fig. 8 The average pea seedling dry weight after 7 days in 1979. (See caption Figure 7).

Similar to the faba beans the 1979 pea average weights were lower than those in 1978. The lower average dry weights at 5°C were mostly due to Century and P1244095.

There were significant differences among faba bean cultivars in conductivity at all temperatures in 1978 and, except at 15°C, the seed moisture levels had a significant effect on conductivity (Figure 9). Generally Ackerperle and Maris Bead had the greatest amount of leachate and, without exception, Erfordia had the lowest. In contrast to 1978, seed leakage in 1979 increased at the 14% level but then dropped at the 21% level (Figure 10). The cultivars Diana and Ackerperle had the greatest amount of leachate in 1979. Temperature did not seem to have much of an effect on the average conductivity values in either year.

In peas the smooth-seeded cultivars Century and Triumph had the greatest conductivity values at each temperature and moisture level in both 1978 (Figure 11) and 1979 (Figure 12). In addition the significant interactions found in both years appear to be due to differences between smooth- and wrinkle-seeded cultivars. As with the faba beans in 1978, the amount of leachate tended to decline at the 14% moisture level in 1978 while the main decline in 1979 was at the 21% level. In peas temperature also had little effect on the amount of leachate.

For faba beans in 1978 the correlation coefficient for the number of normal seedlings and conductivity varied from -0.34 ($P < 0.05$) at 5°C to -0.09 ($P > 0.1$) at 10°C to 0.31 ($P < 0.1$) at 15°C. In 1979 there were significant negative correlations at all temperatures: -0.72, -0.56 and -0.60 ($P < 0.01$) for 5, 10 and 15°C, respectively.

For peas in 1978 significant negative correlations between the number of normal seedlings and conductivity were obtained at all temperatures, being -0.5 ($P < 0.01$), -0.33 ($P < 0.05$) and -0.44 ($P < 0.01$), respectively. In 1979 significant negative correlations were obtained only at 5 and 10°C and these were -0.37 and -0.36 ($P < 0.05$), respectively. The -0.13 value obtained at 15°C was not significant.

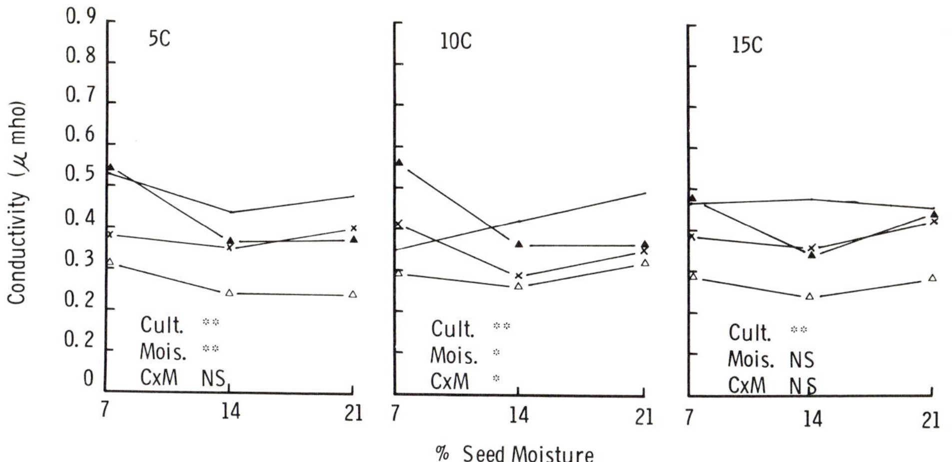

Fig. 9 The 1978 (a) and 1979 (b) average seed leachate conductivity values of the faba bean cultivars Ackerperle (———), Diana (x———x), Erfordia (Δ———Δ) and Maris Bead (▲———▲) soaked in distilled water for 24 h at 5, 10 or 15°C and varying in initial seed moisture. ** and * significant at the 1 and 5% level of probability, respectively. NS: non-significant.

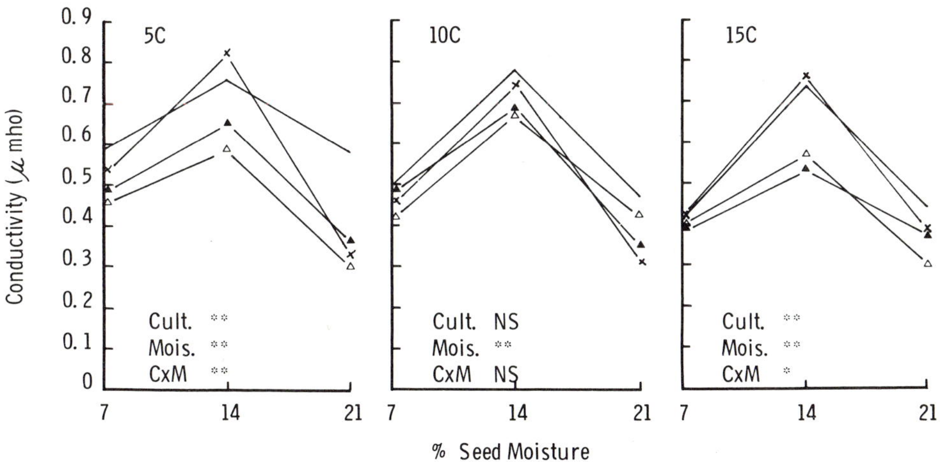

Fig. 10 The 1979 average seed leachate conductivity values for the faba beans. (See caption Figure 9).

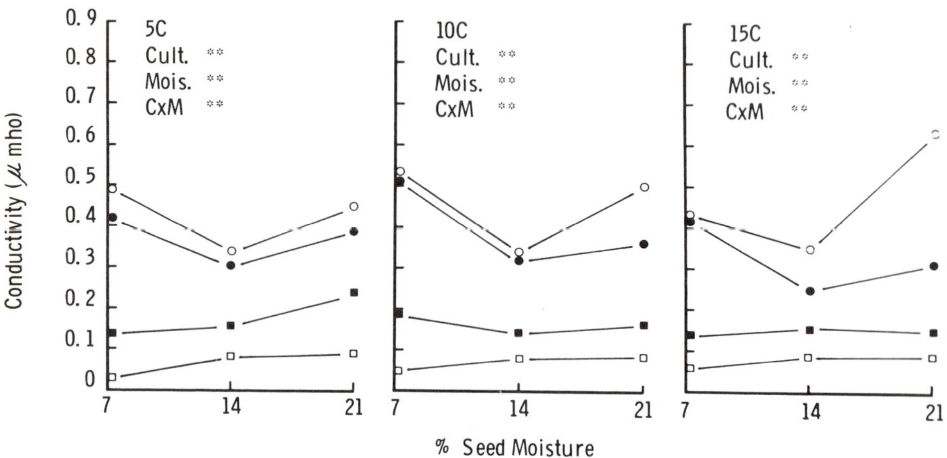

Fig. 11 The 1978 (a) and 1979 (b) average seed leachate conductivity values of the pea cultivars Triumph (o——o), Century (●——●), PI280064 (□——□) and PI244095 (■——■) soaked in distilled water for 24 h at 5, 10 or 15°C and varying in initial seed moisture. ** significant at the 1% level of probability.

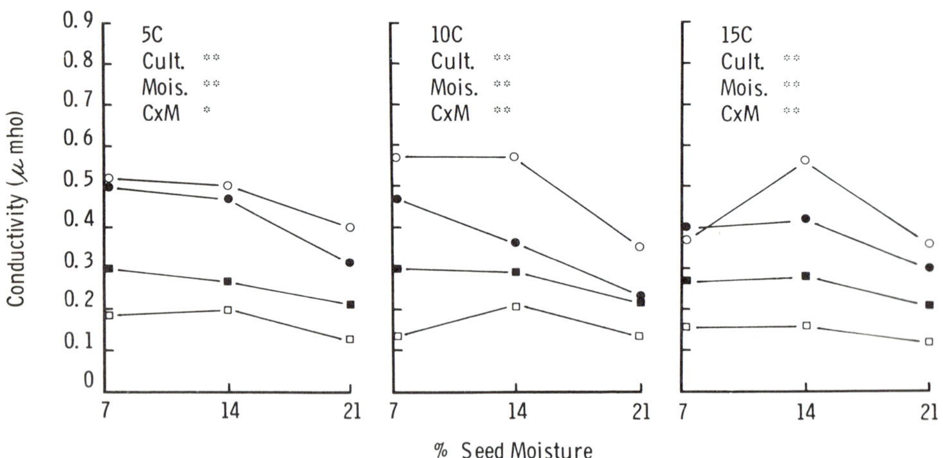

Fig. 12 The average seed leachate conductivity values for the pea cultivars. (See caption Figure 11).

Field experiments

Soil temperatures were already fairly high when seeding started in 1978 (Figure 13a). In 1979 soil temperatures started very low but warmed up quickly after seeding had commenced (Figure 13b).

Dates of seeding had a significant effect on yield of faba beans in 1978 but not in 1979 (Figure 14). Erfordia generally outperformed the other cultivars in both years, while Maris Bead had the lowest yields. However, the highly significant cultivar x date interaction in both years shows the different responses of cultivars to environmental change. The yield of faba beans was greater in 1979, particularly at the third seeding date.

Peas behaved much like faba beans with the first two dates of seeding having higher yields in 1979 than in 1978 (Figure 15). Century consistently gave the highest yields at all seeding dates in both years, while the other cultivars were quite variable in their responses.

Although overall seed moisture levels had no significant effect on yields of peas, there was a significant cultivar x moisture interaction in 1979 (Figure 16). This was primarily caused by the low yield of Triumph at the 14% moisture level which occurred at the second and third seeding dates, but not the first.

In faba beans in both years seeding dates and cultivars had a significant effect on days to flower and there were significant cultivar x date interactions (Figure 17). The number of days to flower decreased as seeding was delayed. Peas behaved much the same with the exception that the differences among cultivars were smaller.

The number of days to swathing in faba beans was smaller in 1979 than in 1978, except for the middle seeding date (Figure 18). Diana was about five days earlier maturing than the others. In 1979 plants from the 14% moisture seed were slightly slower

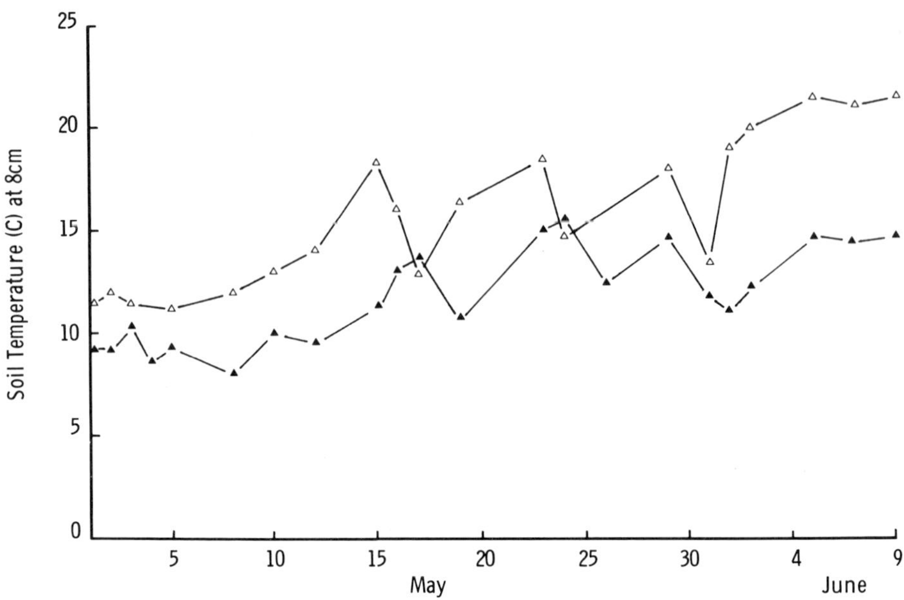

Fig. 13a 1978 soil temperatures at 8 cm depth. △——△ day temperature and ▲——▲ night temperature.

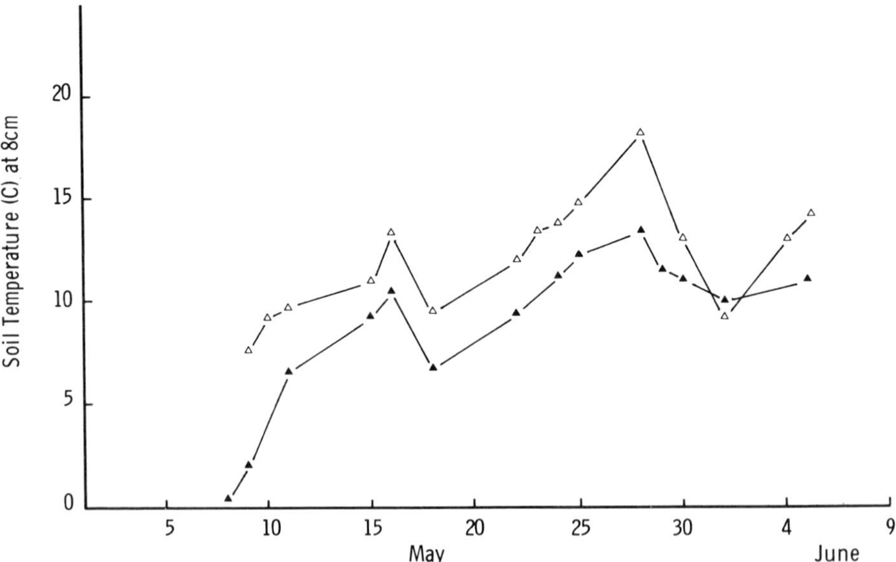

Fig. 13b 1979 soil temperatures.

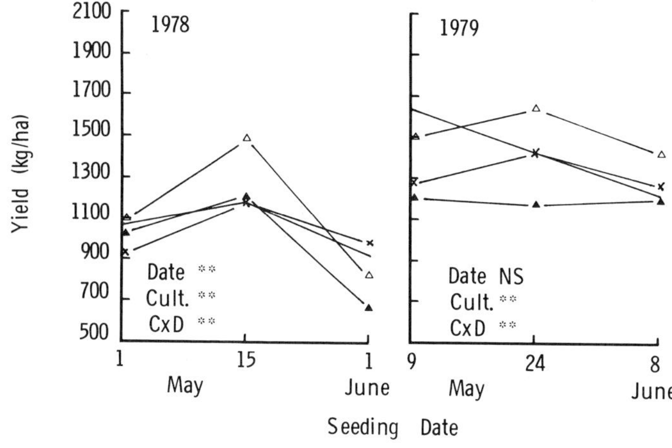

Fig. 14
The average seed yield (kg/ha) of the faba bean cultivars:
Ackerperle (————),
Diana (x————x),
Erfordia (Δ————Δ)
and
Maris Bead (▲————▲).
** significant at the 1% level of probability.
NS: non-significant

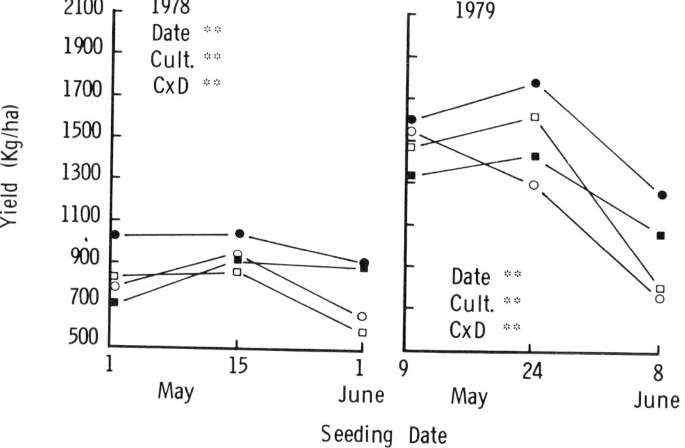

Fig. 15
The average seed yield (kg/ha) of the pea cultivars:
Triumph (o————o),
Century (●————●),
PI280064 (□————□)
and
PI244095 (■————■).
** significant at the 1% level of probability.

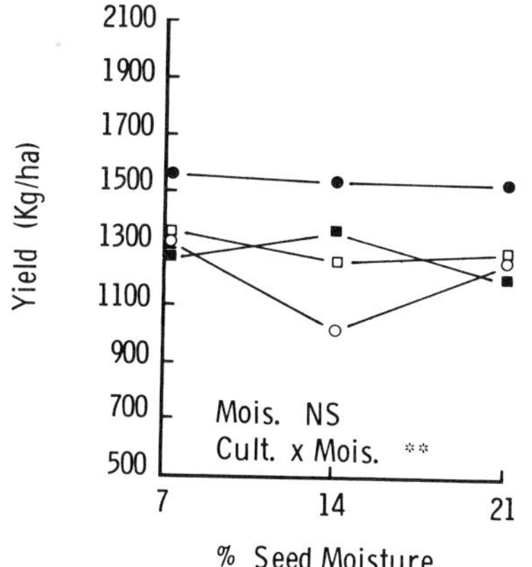

Fig. 16
The average 1979 seed yield of the pea cultivars according to initial seed moisture level. (See Figure 15 for symbols).

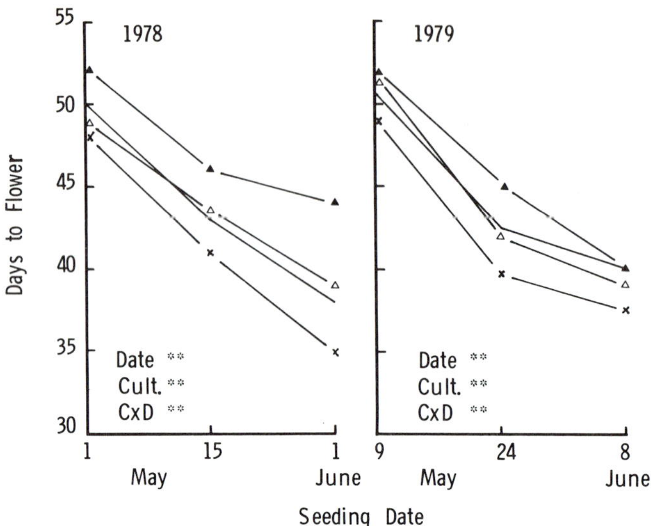

Fig. 17 The average number of days to first flower of the faba bean cultivars Ackerperle (———), Diana (x——x), Erfordia (Δ——Δ) and Maris Bead (▲——▲). ** significant at 1% level of probability.

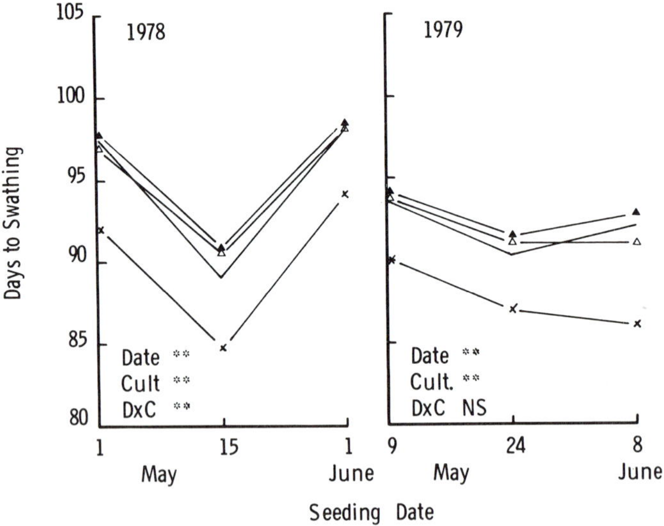

Fig. 18 The average number of days to swathing of the faba beans. (See Figure 17 for symbols).

maturing (Figure 19). No record of days to maturity was kept for peas as they were harvested at varying stages of maturity.

The numbers of faba bean plants per 2 m row did not differ significantly except for the effects of moisture levels in 1979 ($P < 0.01$). The 7, 14 and 21% moisture levels averaged 19.6, 18.1 and 21.1 plants per 2 m of row respectively. In peas the plant numbers generally fell after the first planting date in both years (Figure 20). The number of plants of Triumph was much lower than the others in 1979.

DISCUSSION

Despite the wide differences found in germination, seedling dry weights and conductivity among cultivars in both faba beans and peas, no substantial effects of seed moisture levels on yield and plant numbers could be found. This was true even though there were tremendous differences in both the seedbed environment and seed quality between the two years. Germination percentages were much greater for the seed used for 1978 than that in 1979. Nevertheless, the yields of the 1979 experiments were greater despite the seedbed being much cooler throughout most of the germination and early growth period. The average difference in germination of the 1979 seed versus the 1978 seed was 38% for faba beans and 25% for peas, while the average difference in field plant numbers was only 13.7 and 14.7%, respectively.

Significant negative correlations were found in most cases between the numbers of normal seedlings and conductivity of seed leachates in both species. This is in contrast to Hegarty's (1977) findings in faba beans of $r = 0.27$ and Powell and Matthews (1979) in peas of $r = 0.15$. These values are in agreement however with Scott and Close (1976) who found a correlation coefficient of -0.6 in peas. The varying results do emphasise that seed from different years and locations can give quite different results and also that the temperature used for water imbibition and conductivity might influence the results. Hegarty (1977) and Powell and Matthews (1979) used temperatures of 18 and 20°C respectively,

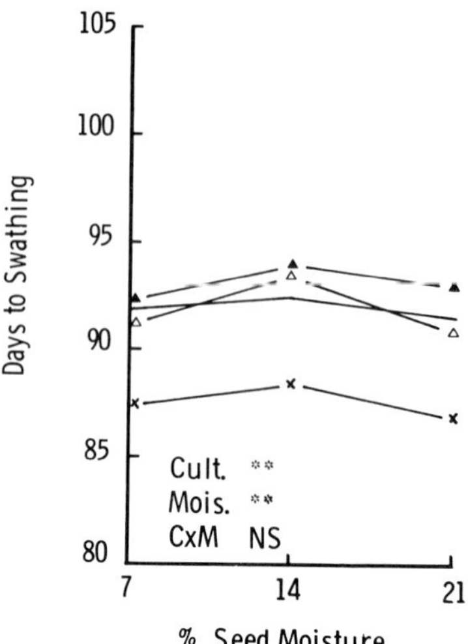

Fig. 19 The 1979 average number of days to swathing of faba beans according to initial seed moisture level. (see Figure 17 for symbols).

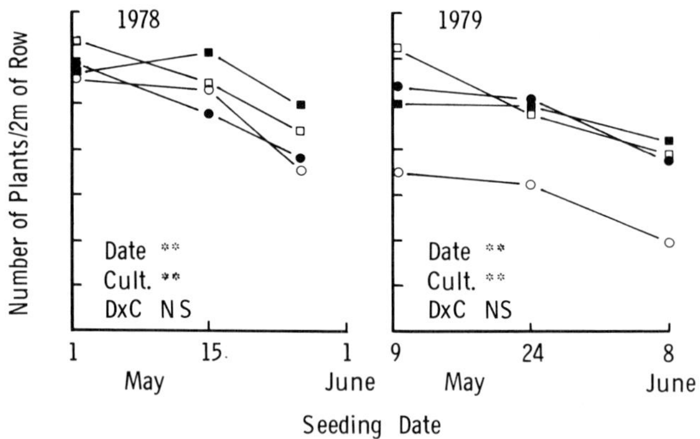

Fig. 20 The average number of pea plants per 2 m row for the cultivars Triumph (o——o), Century (●——●), PI 280064 (□——□) and PI244095 (■——■). ** significant at the 1% level of probability. NS: non-significant.

while the greater negative correlations in this study were generally found at 5°C.

Germination temperatures caused some differences in the seedling dry weights of faba beans and peas in 1978, but the differences were much smaller in 1979. Certainly the different effects of temperatures were not as large as those found by Hardwick (1978) on *Phaseolus vulgaris* and *P. coccineus*. Bramlage et al. (1978) found that the elongation of axes of soya bean embryos that had imbibed water at 14°C or lower was sharply reduced. Thus, while temperature may have a profound effect on the growth of semi-tropical legumes such as soya beans, the effect on temperate legumes is generally minimal.

The conductivity values were much higher in 1979 than 1978 and this was another indication of the poorer quality of the 1979 seed. While the conductivity results here cannot be directly compared with those of Rowland and Gusta (1977), because they are not presented as μ mho/g of seed, the rankings are interesting. In both 1978 and 1979 the rankings of faba bean cultivars for conductivity were different from that of Rowland and Gusta (1977), once more displaying the effect of year on seed quality. Not only are field results confounded by the environment in which they are grown but also by the environment that produced the seed for the trial.

High conductivity values were generally correlated with poor performance in the germination test but they could not be related to field performance in either yield or plant stand. Certainly the relationship was not of the order of $r = -0.72$ reported by Powell and Matthews (1979) in peas or $r = -0.77$ reported by Hegarty (1977) in faba beans.

The one instance where there did seem to be a relationship between the germination and conductivity data with field data was in the 14% moisture faba bean seed in 1979. At this moisture level germination was reduced and conductivity was higher at each temperature. Plant numbers in the field were significantly reduced and days to harvest were increased with the latter presumably resulting from the thinner stand being later maturing.

Perry and Harrison (1970) working on peas found declining conductivity values as temperature increased, but this was not found here in either peas or faba beans. They attributed the increased conductivity values at low temperatures to an increase in the number of peas killed rather than an overall increase in the level of exudation. Their explanation may have some merit as the average number of normal seedlings in this study for all cultivars over both years and all three temperatures were 26.8, 26.3 and 27.2, respectively and the conductivity values were nearly equal.

Powell and Matthews (1978) felt that it is not low temperatures that cause the loss in germination but rather the rapid uptake of water. Low temperatures simply exaggerate damage that is already present by prolonging the period of rapid leakage (Bramlage et al., 1978). It could be, however, that any stress such as dryness and/or warm temperatures in the seedbed prior to emergence could cause germination problems. A possible example of this would be the significantly lower yield of Triumph peas at 14% moisture at the second and third seeding dates in 1979. By this time the soils were above 10°C which was considerably warmer and drier than the 2°C temperature of the first seeding date.

In spite of the tenuous link between germination failure and seed yield, it would be useful for plant breeders to subject potential cultivars to germination tests at 5°C and conductivity tests prior to their release. This multiple seed testing suggested by Scott and Close (1976), if carried out for several years would weed out cultivars like Triumph peas, which have very poor emergence under stress conditions.

The protective advantage of high moisture seed for Western Canadian soils is questionable. These soils, though cool in early May, warm quickly and the seed is in the soil for only ten days to two weeks before emergence. This is in contrast to Europe where six or more weeks may elapse before emergence.

The results here reaffirm the findings of Rowland and Gusta (1977) that wrinkle-seed pea cultivars do not necessarily have greater seed leakage and poorer germination than smooth-seeded types. The smooth-seeded Triumph generally was the lowest yielding, had the fewest normal seedlings, had the greatest conductivity and had the poorest field emergence of the four pea cultivars used. This poor performance cannot be attributed to Triumph's large seed size as P1244095 has an even greater seed size than Triumph.

ACKNOWLEDGEMENTS

The germination and conductivity tests were carried out in a highly efficient manner by Mr. W. Dale and his staff at the Plant Products Seed Inspection Department of Agriculture Canada in Saskatoon.

REFERENCES

Bramlage, W.J., Leopold, C.A. and Parrish, D.J. 1978. Chilling stress to soybeans during imbibition. Plant Physiology 61, 525-529

Hardwick, R.C. 1978. Effects of low seedbed temperatures on seedling weight. Acta Horticulturae 83, 201-204

Hegarty, T.W. 1977. Seed vigour in field beans *(Vicia faba* L.) and its influence on plant stand. Journal of Agricultural Science, (Cambridge) 88, 169-173

Hobbs, P.R. and Obendorf, R.L. 1972. Interaction of initial seed moisture and imbibitional temperature on germination and productivity of soybean. Crop Science 12, 664-667

Perry, D.A. and Harrison, J.G. 1970. The deleterious effect of water and low temperature on germination of pea seed. Journal of Experimental Botany 21, 504-512

Powell, A.A. and Matthews, S. 1978. The damaging effect of water on dry pea embryos during imbibition. Journal of Experimental Botany 29, 1215-1229

Powell, A.A. and Matthews, S. 1979. The influence of testa condition on the imbibition and vigour of pea seeds. Journal of Experimental Botany 30, 193-194

Rowland, G.G. 1978. Effects of planting and swathing dates on yield, quality and other characters of faba beans *(Vicia faba)* in central Saskatchewan. Canadian Journal of Plant Science 58, 1-6

Rowland, G.G. and Gusta, L.V. 1977. Effects of soaking, seed moisture content, temperature and seed leakage on germination of faba beans *(Vicia faba)* and peas *(Pisum sativum)*. Canadian Journal of Plant Science 57, 401-406

Scott, D.J. and Close, R.C. 1976. An assessment of seed factors affecting field emergence of garden pea seed lots. Seed Science & Technology 4, 287-300

DISCUSSION

R.B. Austin *(UK)*

Have you any idea which factors were responsible for the differences in seed quality between the two years? What were the environmental factors during the growing of the seed crop? For instance, were they grown in the same location?

G.G. Rowland *(Canada)*

Yes, they were all grown in the same location; the same area of the field. In 1978 they were slightly later maturing and frost was not so severe as to affect the quality of the seed very much.

Kleinhout *(Netherlands)*

What was the reason for the low yield of faba beans in Saskatoon?

G.G. Rowland

Rainfall. Our average moisture fall, between snow and rain in a year, is only about 15 - 16 inches. This is much lower than in many other areas of Western Canada.

Kleinhout

In these other areas of Western Canada where there is more moisture, do you have higher yields?

G.G. Rowland

Yes. About 50 - 100 km north of Saskatoon there are much higher yields of faba beans. In Saskatchewan, most of the beans are now grown under irrigation. The yields range from 3½ to slightly over 4 t/ha.

Kleinhout

Another thing is the seed moisture referred to in your Tables - were they treated seeds?

G.G. Rowland

No, the seed was not treated.

Kleinhout

They were not treated but they were pre-treated with water?

G.G. Rowland

Yes, pre-treated with water, but there was no fungicide or anything else applied to the seed.

Kleinhout

If you sowed seed at a depth of 7 cm in the soil, would you have any problems with germination?

G.G. Rowland

We did this one year and this is what started all of this work. We noted certain cultivars with particularly poor emergence. The pea cultivar Triumph consistently had poor germination year after year, especially under stress conditions.

R. Thompson *(UK)*

Did you make any assessment of the amount of hollow-heart in the peas in particular? You say that the weather is dry and can be hot, these are conditions which favour development of hollow-heart and this could account for the differences.

G.G. Rowland

We had looked at this before and found no evidence of hollow-heart, however, we did not look with the seed that was used in the 1979 experiment thus it is possible that something like that occurred.

GRAIN YIELD STABILITY OF FIELD BEAN (*Vicia faba* L.) IN NORTH-EASTERN ITALY

U. Ziliotto, G. Mosca and L. Toniolo
Institute of Agronomy, University of Padua,
Via Gradenigo 6, 35100 Padua, Italy

ABSTRACT

*During the three year period, 1976 - 1978, a number of agronomic trials were made with Italian and foreign cultivars of field beans (*Vicia faba *L. var.* minor *Beck. and* V. faba *L. var.* equina *Pers.) in the Venetian plains, with particular reference to the influence of sowing date, both autumn and spring, on growth and yield.*

The average productivity and stability in relation to the sowing season were examined for two groups of cultivars, one group comprising 10 trials and the other group 7 trials.

The cultivars which proved to be the most productive of the first group were Ascott, Diana, Skladia kleine, Maxime, Manfredini and Ackerperle, and the most productive in the second group was Wierboon CB.

In the first group, which was composed of 12 cultivars, eight showed a significantly high stability index (P $<$0.05), and among these Primperle was to be the most stable while Manfredini gave the highest b value (slope for regression of variety yield on environment mean).

The reason for the non-significance of b as regards L.72, Vesuvio, Herra and Pavane is attributed to the autumn behaviour of the first two varieties and to the susceptibility to cold of the second two, which thus makes these more suitable for spring sowing.

Within the second group, Felix proved to be the most stable cultivar.

INTRODUCTION

There are various reasons why, in the northern flat lands of the Po valley, the cultivation of field beans may be of increasing interest. Agricultural practice in this region has resulted in the total disappearance of local traditional varieties and our knowledge of the performance of the newest varieties in different parts of this region is limited.

Experiments involving spring and autumn sowing of Italian and foreign varieties have been carried out to improve our understanding of the performance of the species in this region. Particular attention in this paper is paid to yield and its relative stability.

MATERIAL AND METHODS

The trials took place over a three year period, from 1976 - 1978 at the Experimental Farm of the Agricultural Department (Legnaro, Province of Padua). The clay soils were neutral or alkaline in pH reaction and were very rich in limestone and poor in organic matter. They contained medium levels of N and K_2O and medium to high levels of available P_2O_5.

The autumn sowings took place from early October to the end of November, while the spring sowings were made from early February to early April. All the trials were done in randomised blocks with 3 or 4 replicates and 8 m^2 plot sizes. Row widths were 25 cm and the nominal plant density was 40 plants/m^2. A base fertiliser of 150 kg/ha of P_2O_5 and 100 kg/ha of K_2O was applied. The varieties common to the same trials were arranged together, thus obtaining two groups of cultivars (Table 1).

Each experiment was analysed by the methods described by Finlay and Wilkinson (1963). Within each group, the differences between cultivars were tested at the 0.05 probability level with Duncan's new multiple range test. Given a significant effect in each group for the genotype x environment interaction, the

TABLE 1

DETAILS OF THE NUMBERS OF CULTIVARS DIVIDED ACCORDING TO VARIOUS 'ENVIRONMENTS' USED IN THE TWO SETS OF EXPERIMENTS

Groups of cultivars	No. of cultivars	'Environments'					Total
		Spring sown			Autumn sown		
		1976	1977	1978	1976	1977	
1	12	2	4	-	-	4	10
2	5	-	2	2	1	2	7

regression of cultivar yield values against environment means was used to determine differences in stability.

In 1976, rainfall was lower and temperatures were much higher than average (Table 2). In 1976 - 1977 and 1977 - 1978, rainfall was higher than average. The temperature for the winter season was consistently higher than normal, but that in the spring was near average for that time of year.

RESULTS

First group of cultivars

The average grain yield obtained from 12 cultivars (Table 3) in 10 environments was 2.19 t/ha of dry matter. The highest yielding cultivars were Ascott, Diana, Skladia kleine, Maxime, Manfredini and Ackerperle, ranging from 2.24 to 2.72 t/ha of dry matter. The least productive, but not necessarily significantly different from the preceding cultivars, were Minor, Primperle, Pavane and Herra (range 1.61 - 2.01 t/ha dry matter).

In general, delayed sowing reduced the average yield. Eight cultivars showed highly significant stability indices. Primperle gave a value for b of less than 1 and Manfredini gave a value exceeding 1 (Figure 1). More specifically, the grain yield of Primperle equalled that of the others in poor environments while in better ones it was signficantly less productive.

TABLE 2

METEOROLOGICAL DATA FROM LEGNARO (PADUA - ITALY) DURING THE EXPERIMENTAL PERIODS

Month	Average temperature (^{o}C)				Rainfall (mm)			
	Monthly temperature			15 year average	Monthly rainfall			15 year average
	1976	1976-77	1977-78		1976	1976-77	1977-78	
October	-	13.5	13.4	11.7	-	209.8	44.9	61.8
November	-	8.5	7.2	6.0	-	77.5	62.2	77.9
December	-	3.0	2.5	1.6	-	103.7	50.3	59.4
January	-	3.9	3.3	1.4	-	118.8	131.8	68.7
February	5.3	6.1	3.3	4.5	73.5	51.3	107.1	64.4
March	5.4	9.3	7.9	7.2	17.7	91.1	41.0	59.7
April	11.6	10.4	9.7	10.7	49.6	78.6	126.4	66.8
May	16.9	15.3	13.5	15.4	19.4	96.8	83.1	70.5
June	21.8	19.7	20.4	19.5	6.5	39.0	51.2	74.5
July	24.3	21.6	20.9	21.5	116.2	133.1	46.4	85.6
Total or mean	(14.2)	11.1	10.2	10.0	(282.9)	999.7	744.4	689.3

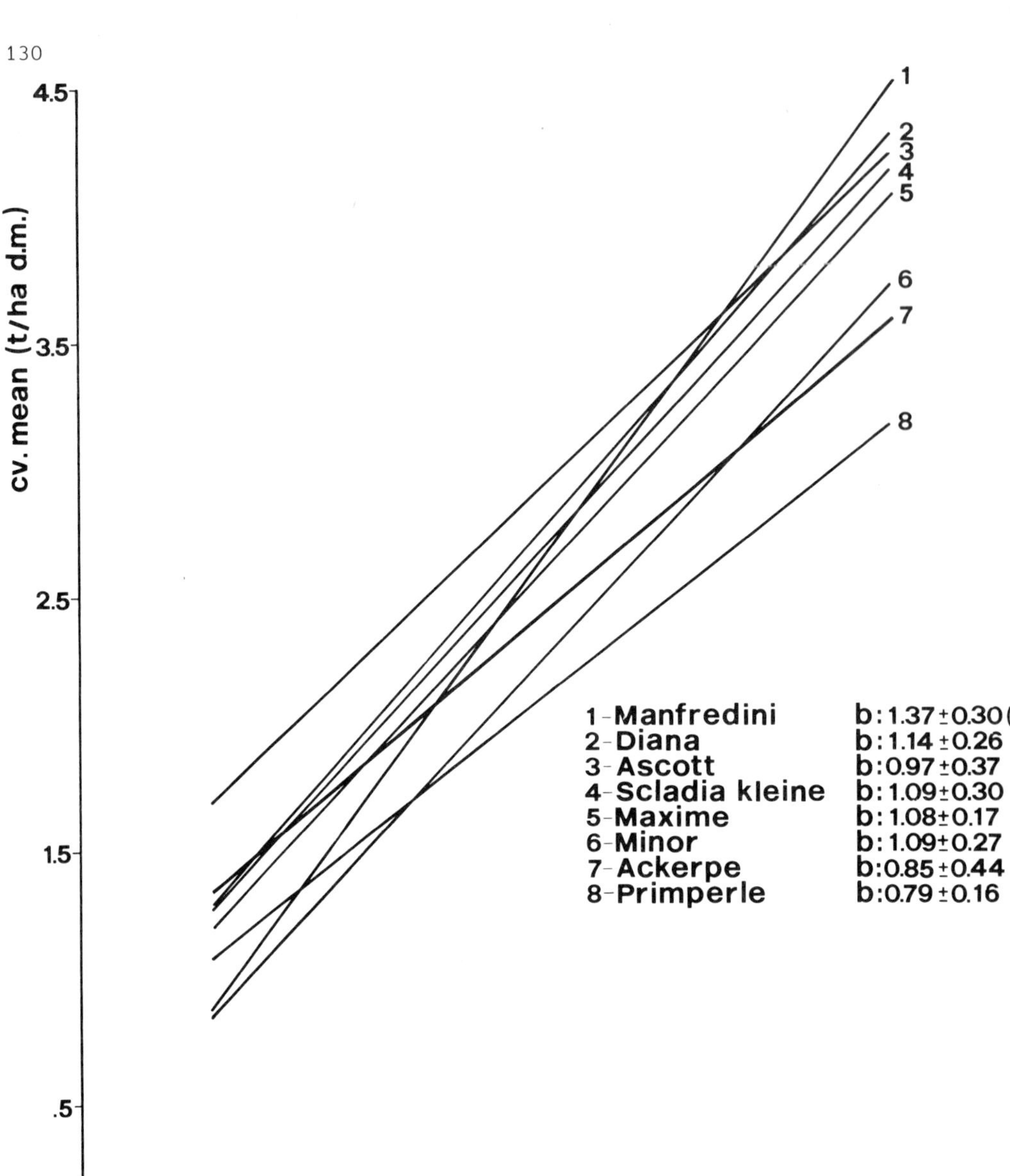

Fig. 1: Regression lines of eight (lst group) cultivar means on year/sowing date means.
e_1 = 15/4/1977; e_2 = 5/4/1977; e_3 = 29/3/1976; e_4 = 14/3/1976; e_5 = 18/3/1977; e_6 = 29/11/1977; e_7 = 14/11/1977; e_8 = 2/3/1977; e_9 = 31/10/1977; e_{10} = 18/10/1977.
(*) - 95% confidence level.

TABLE 3

YIELD (t/ha) AND VALUES OF b FOR THE FIRST GROUP OF CULTIVARS

Cultivars	Country of origin	Grain yield (1) (t/ha DM)	'b'
Ascott	France	2.72 a	0.97***
Diana	West Germany	2.51 ab	1.14***
Skladia kleine	West Germany	2.45 ab	1.09***
Maxime	Belgium	2.36 ac	1.08***
Manfredini	Italy	2.35 ac	1.37***
Ackerperle	West Germany	2.24 ac	0.85**
Vesuvio	Italy	2.18 bc	1.34 NS
L.72	Italy	2.13 bc	1.47 NS
Minor	Belgium	2.01 bd	1.09***
Primperle	France	1.92 cd	0.79***
Pavane	France	1.85 cd	0.46 NS
Herra	West Germany	1.61 d	0.36 NS
Mean		2.19	1.00

(1) Cultivar means followed by the same letter do not differ at the 0.05 probability level.

*** = significance at 0.001P; ** = significance at 0.01P; NS = not significant

The coefficients of the other six varieties did not differ significantly from 1.0 indicating that their yield increased in direct proportion to the productive capacity of the environment. The regression for Vesuvio, L.72, Herra and Pavane did not reach significance, which indicated their extreme variability. The cause of this variability, was their differing performance when sown in autumn or spring time (Figure 2).
It may be concluded that:

1. Cultivars L.72 and Vesuvio both gave high yields in good environmental conditions (= autumn sowings) but, when sown in the spring, yield was reduced and declined sharply as the environment became progressively unfavourable i.e. when sowing time was delayed.

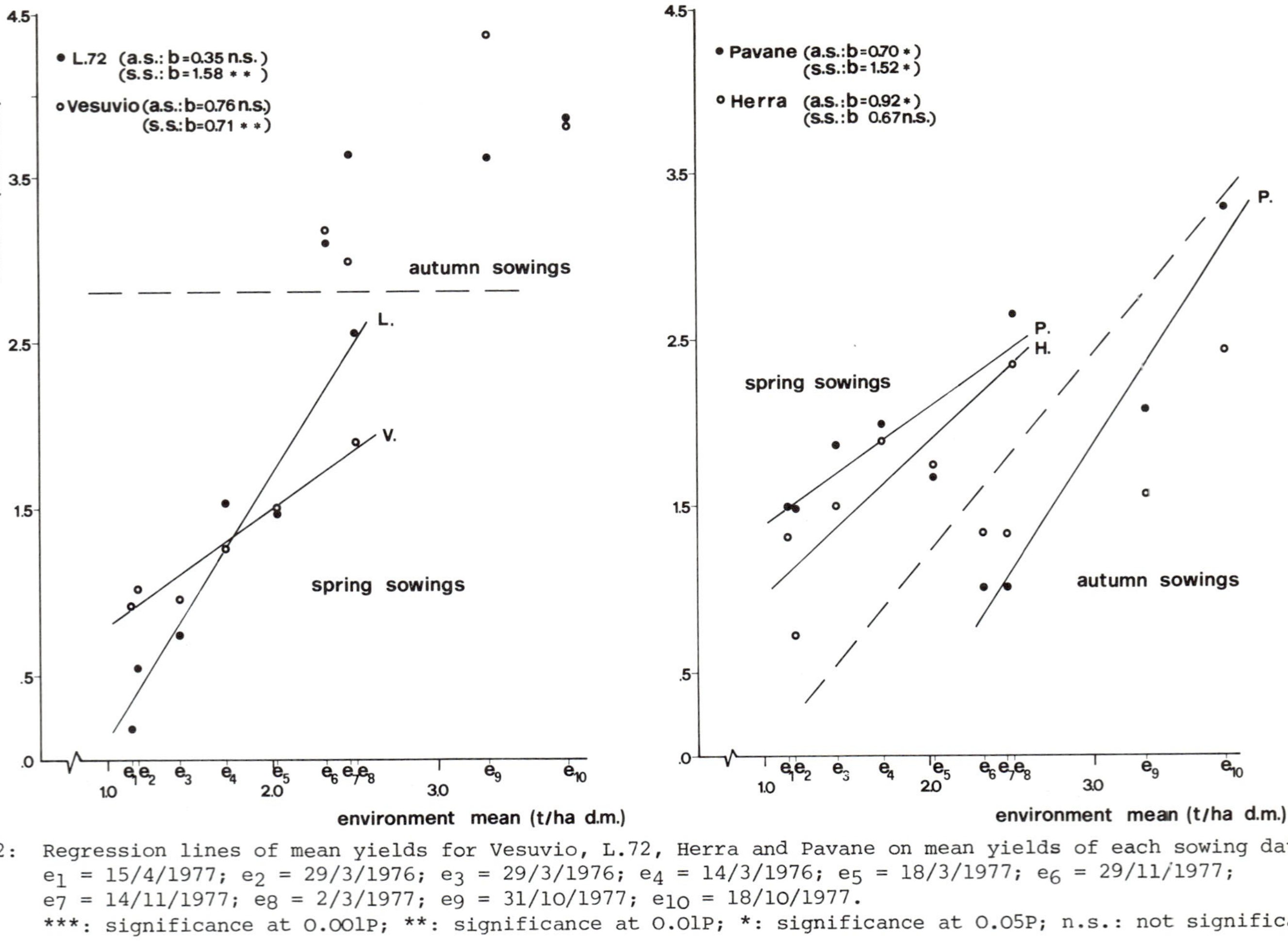

Fig. 2: Regression lines of mean yields for Vesuvio, L.72, Herra and Pavane on mean yields of each sowing date. e_1 = 15/4/1977; e_2 = 29/3/1976; e_3 = 29/3/1976; e_4 = 14/3/1976; e_5 = 18/3/1977; e_6 = 29/11/1977; e_7 = 14/11/1977; e_8 = 2/3/1977; e_9 = 31/10/1977; e_{10} = 18/10/1977.
***: significance at 0.001P; **: significance at 0.01P; *: significance at 0.05P; n.s.: not significant.

2. Cultivars Herra and Pavane gave similar yields from either spring or autumn sowings. Moreover, the responses to delay in sowing with these cultivars were similar in terms of reduced yields for both spring and autumn sowing.

Generalising, it seems that cultivars L.72 and Vesuvio are winter types, while Herra and Pavane proved to be susceptible to cold climates for which reason they are more suitable for spring sowing.

Second group of cultivars

The average grain yield obtained from the second group of cultivars (Table 4), in 7 different environments, was 2.31 t/ha dry matter.

It is worth noting that Wierboon CB was the most productive followed by Minica, Kristall and Maris Bead. The lowest yielding was Felix whose yield was not significantly different from that of Maris Bead. Minica was the cultivar with the greatest variation around its regression line (b = 0.99 ± 0.66; $P < 0.05$) (Figure 3). On the other hand, the value of b for Felix was 0.74 ± 0.22, making it the most stable, whilst Wierboon CB and Maris Bead were least stable.

In the second group, the stability increased as yield decreased. However, the rank order of the cultivars remained constant when they were grown in different environmental conditions.

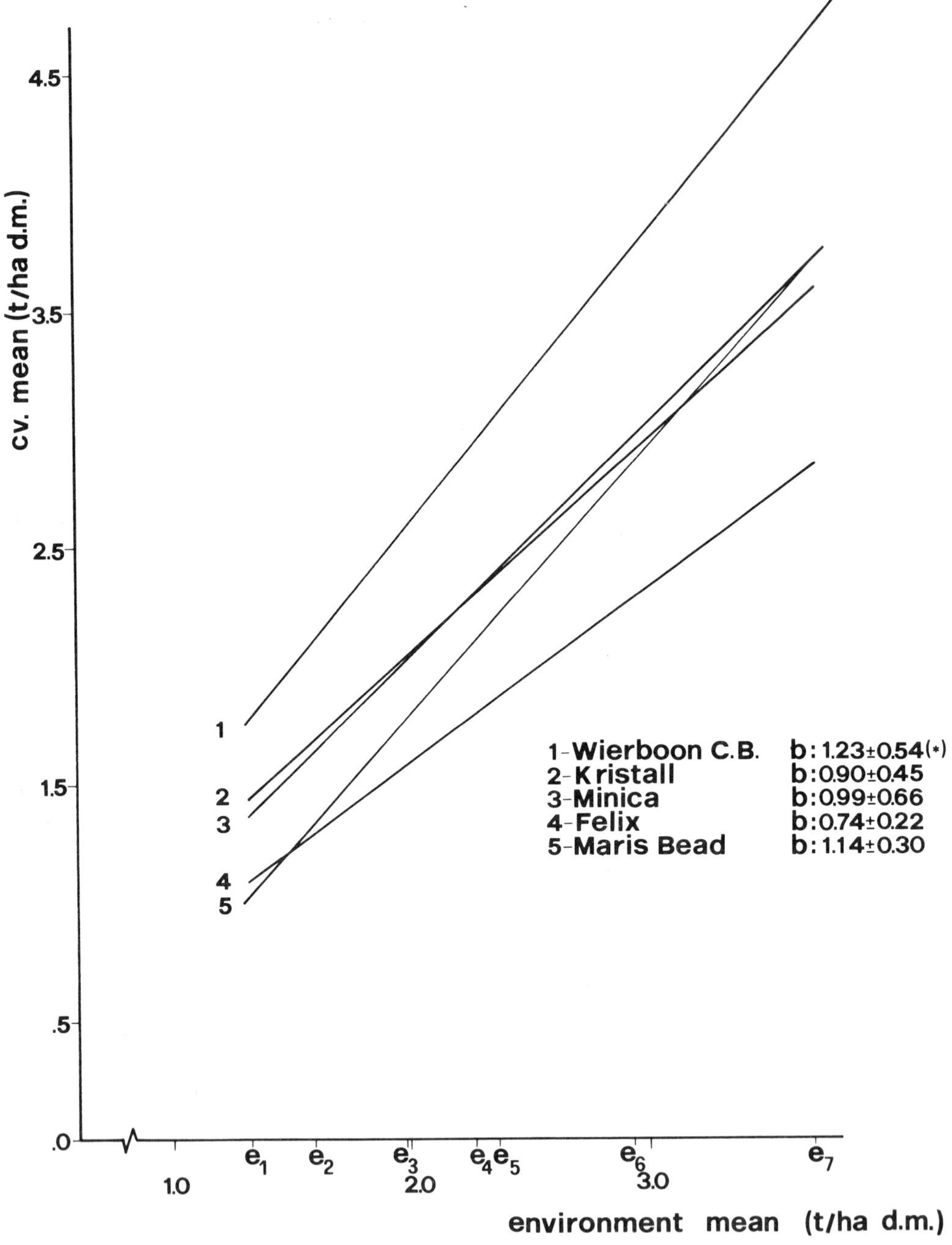

Fig. 3: Regression lines of five (2nd group) cultivar means on year/sowing date means.
e_1 = 29/3/1978; e_2 = 9/3/1977; e_3 = 29/3/1978; e_4 = 9/3/1977; e_5 = 6/10/1977; e_6 = 6/10/1977; e_7 = 30/11/1976.
(*) - 95% confidence limits.

TABLE 4

YIELD (t/ha) AND VALUES OF b FOR THE SECOND GROUP OF CULTIVARS

Cultivars	Country of origin	Grain yield (1) (t/ha DM)	'b'
Wierboon CB	Netherlands	2.92 a	1.23**
Minica	Netherlands	2.34 b	0.99*
Kristall	West Germany	2.33 b	0.90**
Maris Bead	United Kingdom	2.13 bc	1.14***
Felix	Netherlands	1.82 c	0.74***
Mean		2.31	1.00

(1) Cultivar means followed by the same letter do not differ at the 0.05 probability level.

*** = significance at 0.001P; ** = significance at 0.01P: * = significance at 0.05P.

CONCLUSIONS

On the basis of the results obtained we can draw the following conclusions:

1. As already observed by Ziliotto and Toniolo (1979), the highest yields (and those which did not differ significantly) in descending order were given by Ascott, Diana, Skladia kleine, Maxime, Manfredini and Ackerperle for the first group, and Wierboon CB for the second.

2. The varieties which showed a high level of stability were Primperle and Felix.

3. Cultivars L.72 and Vesuvio were winter types because only with autumn sowing were they likely to give high yields.

4. Cultivars Herra and Pavane were sensitive to cold climates and were, therefore, more adapted for spring sowing.

REFERENCES

Finlay, K.W. and Wilkinson, G.N., 1963. The analysis of adaptation in a plant-breeding programme. Australian J. Agric. Res., 14, 742-754.

Ziliotto, U. and Toniolo, L., 1979. Field trials on time of sowing field beans (*Vicia faba* var. *minor*) and broad beans (*Vicia faba* var. *major*) at Padova in 1976 and 1977. In D.A. Bond, G.T. Scarascia-Mugnozza and M.H. Poulsen (Eds.): Some current research on *Vicia faba* in Western Europe, 189-213. ECSC-EEC-EAEC, Brussels - Luxemburg.

DISCUSSION

M.H. Poulsen *(Denmark)*

Have you observed any problems with seed set in the German or Dutch varieties? You mentioned seven locations; were they spread all over Italy or were they only spread around Padua or the northern part of Italy? Perhaps you do not have the same kind of problems as in southern Italy with the western and northern European materials.

G. Mosca *(Italy)*

With regard to the first question the location only applies to the Padua area and not other environments.

As for your second question, in southern Italy the environmental conditions are very different from ours and I do not have any results for southern Italy.

J. Kleinhout *(Netherlands)*

Did you observe any relationship between stability and particular characters of the varieties concerned?

G. Mosca

We have made observations on other characters but are not reporting them in this paper.

D.A. Bond *(UK)*

Wierboon and Minica were the unstable varieties and they are large seeded. Is there a relationship between seed size and stability? Are the large seeded varieties unstable and the small seeded varieties more stable? What about Manfredini, for example, what is the seed size for this variety?

G. Mosca

The weight of the seed for Manfredini is about 700 mg.

D.A. Bond

That is smaller than Minica and Wierboon. But then there are the small seeded varieties like Herra and Primperle. Can you say if there is an association between stability and smaller seed size?

G. Mosca

I do not know.

R.B. Austin *(UK)*

Did you measure the harvest index of your varieties, and do you know if there is any association between harvest index and stability?

G. Mosca

It was not measured but it may well be similar to your own values.

D.A. Lawes *(UK)*

Thank you again, Dr. Mosca.

FACTORIAL ANALYSIS OF YIELD COMPONENTS IN *Vicia faba*

J-I. Cubero and A. Martín

Departmento de Genética,
Escuela Técnica Superior de Ingenieros Agrónomos,
Córdoba, Spain.

ABSTRACT

Factorial analysis has been used to study the yield components in faba beans. Different results were obtained depending on the kind of material studied. The differences were due to the different genetic architecture of these materials. It has been shown to be important to take account of the genetics of the characters included in the study to explain the results.

Application of this method to a set of F_2*s showed that the most important characters influencing yield, were seeds/plant, pods/node and pods/flower.*

INTRODUCTION

Yield can be analysed in various ways for example by multiple regression analysis, factorial analysis, using lines with similar genetic background but differing in one or several components of yield, or by using physiological techniques etc. A common problem of all these techniques is the selection of material; if the material to be analysed is composed of cultivars or lines, the yield components constitute 'blocks' because the corresponding genotypes have been fixed by selection. The use of F_1 plants by no means solves the problem, because of the existence of heterosis in some characters but not in others (in faba beans, between a third and half of the total number of characters studied show heterosis to a greater or lesser extent). Finally, new combinations of characters will result from the use of F_2s, but it is not possible to use parental lines differing in all possible characteristics, moreover many genotypes will show a high degree of heterozygosity, giving the same problems as those mentioned in the case of the F_1s.

One possible solution is to select lines (F_5 - F_6) derived from F_2 but, even in this case, the results will be a function of the parental structure; if both parents were self-fertile or similar in leaf shape, then leaf shape and self-fertility will probably not be important characteristics defining yield.

To show the importance of the selection of material and, at the same time, to examine the yield components of faba beans, we have studied a set of very different pure lines, the 21 hybrids produced from these lines and several F_2s.

MATERIAL AND METHODS

The pure lines have been described in earlier work; there were two *V. faba major* lines, one *equina*, two *minor* and two *paucijuga* lines. The following characters were recorded:

(1) seeds/pod
(2) pods/node
(3) seeds/plant
(4) pods/plant
(5) pod thickness/width
(6) seed weight
(7) seed length
(8) seed thickness/length
(9) seed thickness/width
(10) first node with pods
(11) seeds/ovule
(12) ovules/ovary
(13) flowers/node
(14) pods/flower
(15) leaves/plant
(16) leaflet length
(17) leaflet width
(18) leaflet area
(19) leaflet density
(20) rachis length
(21) leaflets/leaf
(22) leaflet width/length
(23) plant height
(24) days to flower
(25) number of branches

The same characters were also recorded for the 21 F_1s produced. A total of seven different F_2s were studied. The parental lines were VF 164, VF 165 *(major)*, VF 166 *(equina-minor)*, VF 171 and VF 172 *(paucijuga)*. Because of the number of F_2 plants to be studied, for these only eleven characters were recorded i.e. those numbered (1), (2), (3), (4), (6), (11), (12), (13), (14) and (24) listed above, as well as the number of broomrapes per plant (broomrape attack was very slight, but was included as a potential factor affecting yield).

The principal component analysis was performed according to the method described by Kendall (1972). The analysis was made in the following way: for each line, F_1 or F_2 plant, a single value was obtained by means of a linear combination of the natural characteristics, provided by the eigenvectors. Using the new values as independent variates, the correlation coefficients with the corresponding yields were obtained, as well as the regression lines. Main components of the eigenvector correlated to yield were then identified.

RESULTS AND DISCUSSION

For the sake of simplicity the results are presented as summaries with only a few figures to illustrate the method.

TABLE 1

PRINCIPAL COMPONENT ANALYSIS, PURE LINES

Vector	v1	v2	v3
% of variation explained	40.5	28.0	14.9
% of variation accumulated	40.5	68.5	83.4
Correlation with yield	-0.67**	-0.01	-0.72**
Most important characters and direction of influence on yield.	(14) Pods/fl. (-)		(10) First/pod (-)
	(11) Seeds/ov. (-)		(4) Pods/pl. (+)
	(17) Leaf. wd. (+)		(3) Seeds/pl. (+)
	(23) Height (+)		(15) Leav./pl. (+)
	(12) Ovules/ov. (+)		(5) Pod t/w (-)

(a) Pure lines analysis

Table 1 shows that the transformed variables corresponding to vectors 1 and 3 showed a high and significant correlation with yield. The five most important characters determining the first and the third vectors, and hence yield, are also given, with a sign indicating the direction of influence.

It is surprising to find that the transformation of flowers on pods and that of the ovules on seeds showed a negative influence on yield, but this result is a logical consequence of the material used in this study: pure lines with the highest fertility were accessions 171 and 172 *(paucijuga)* and were the worst yielders.

However, it is not worth discussing the results further because the analysis with pure lines was not very informative as can be deduced from the regression for yield/first vector showed in Figure 1.

(b) F_1 analysis

Table 2 shows that only the first vector is strongly correlated with yield. The absence of both the number of

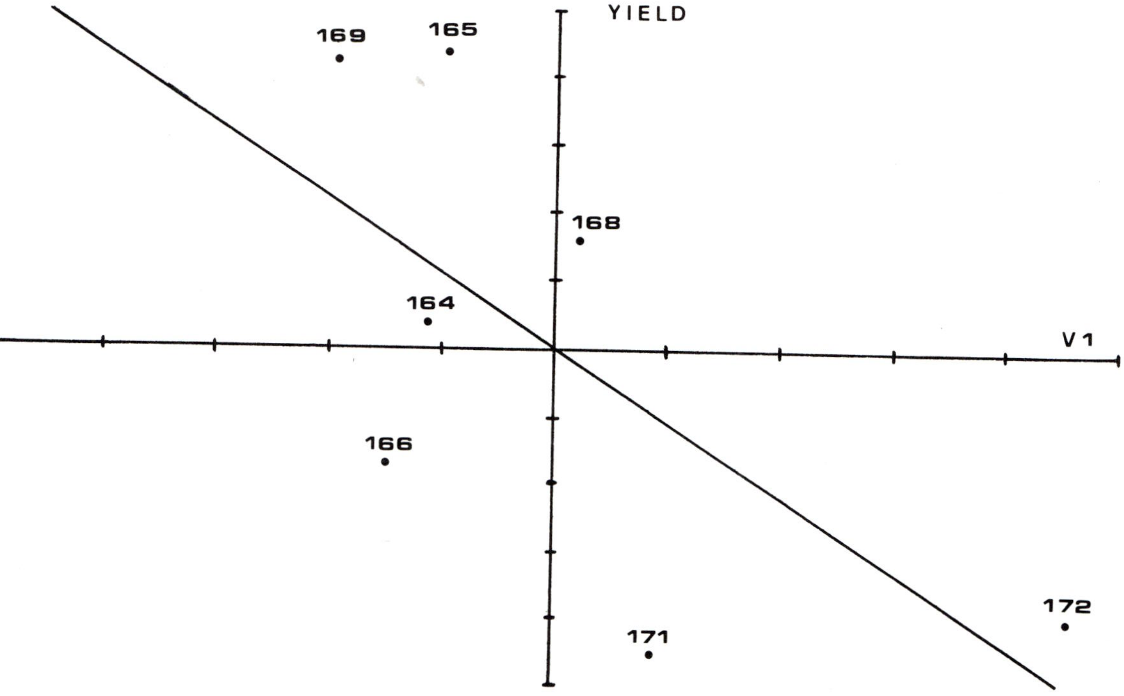

Fig. 1 Principal component analysis for the pure lines.

TABLE 2

PRINCIPAL COMPONENT ANALYSIS, F_1 GENERATION

Vector	v1	v2	v3
% of variation explained	39.4	31.1	9.5
% of variation accumulated	39.4	70.5	80.0
Correlation with yield	0.71**	0.32	-0.18

↓ (v1)

Most important characters and sense of influence on yield		
(13)	Flowers/node	(+)
(17)	Leaflet w.	(+)
(18)	Leaflet area	(+)
(23)	Height	(+)
(3)	Seeds/plant	(-)

pods/flower and of seeds/ovule (two characters measuring self-fertility) is not surprising because hybrids are self-fertile. Even though different F_2s differ in their degree of self-fertility (Martín, 1976; Cubero and Martín, 1978), the high values generally shown by this character result in its absence among the most important factors. This is the reason that flowers/node was the most important character influencing yield of F_1s, for if the degree of self-fertility was high then a larger number of flowers per node would result in greater numbers of pods.

Seeds/plant was negatively related to yield because small seed size is partially dominant over large, and a large number of seeds/pod over a small number (Cubero and Martín, 1978). Thus, *paucijuga* accessions will produce F_1s with many small seeds but with faba leaves and habit (most of *paucijuga* vegetative characters are recessive), which explains the characters and signs listed in Table 2.

These results indicate some of the difficulties of studying yield components even using genetically buffered materials such as F_1s. Figure 2 shows the correlation between yield and the characteristics transformed by the first vector. The vl negative values correspond to the *paucijuga* hybrids, and the maximum positive values to those of *equina*. The best yielders were not those with the largest seeds. This result suggests that there was an optimum for seed size in relation to yield.

(c) F_2 analysis

The results of the F_2 analysis are summarised in Table 3, and Figure 3, for the sake of simplicity, shows only one of the cases (the other F_2s produced very similar distributions). The loss of several F_2s including *major* lines was unfortunate, however the characteristics of the parental lines used in our study show a wide range. Only the first vector, which explained the maximum amount of variation, was correlated (ranging from 0.77 to 0.88) with yield.

Table 3 shows the five most important characteristics in relation to yield for each F_2. Fewer characteristics were examined than for the pure lines and F_1s, but those chosen were more directly related to yield.

It can easily be seen in Table 3 that the characters mainly influencing yield were in order of importance:

(1) seeds/plant
(2) pods/plant
(3) pods/node
(4) pods/flower

The pattern was very similar in all the F_2 plants studied. As seeds/plant and pods/plant were strongly correlated both at the phenotypic and the genotypic levels, and the same was true for pods/node and pods/flower (Martín, 1976), therefore the number of characters necessary for effective selection is reduced to only two i.e. pods/plant and pods/node which are simpler to deal with than seeds/plant or pods/flower.

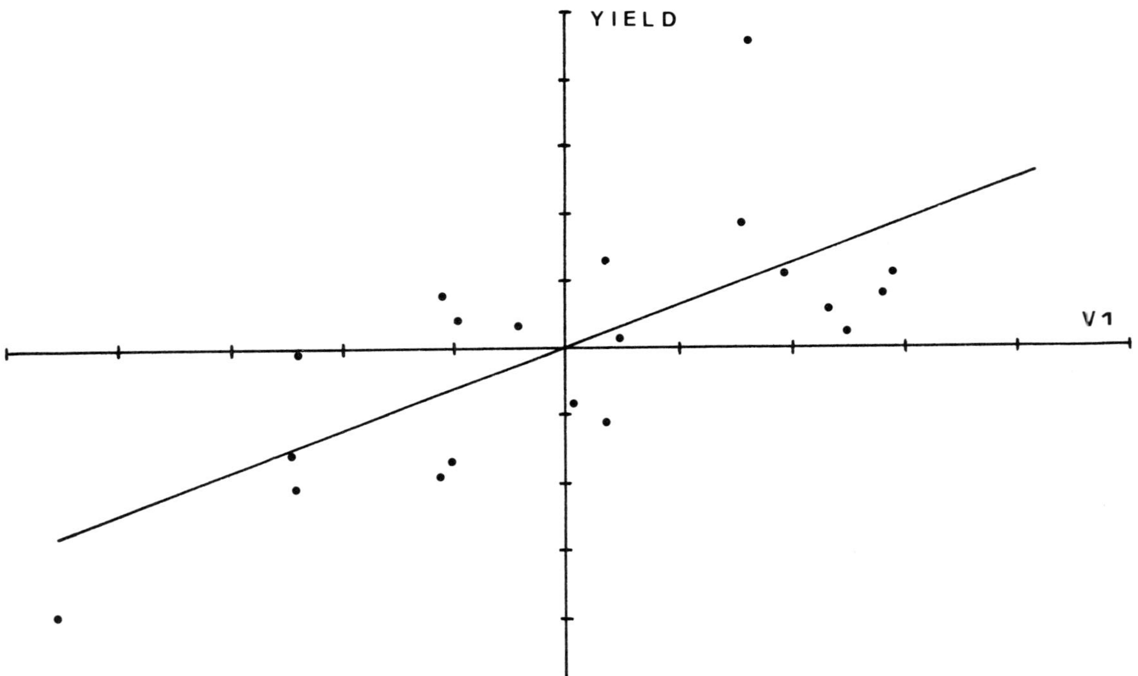

Fig. 2 Principal component analysis for the F_1s.

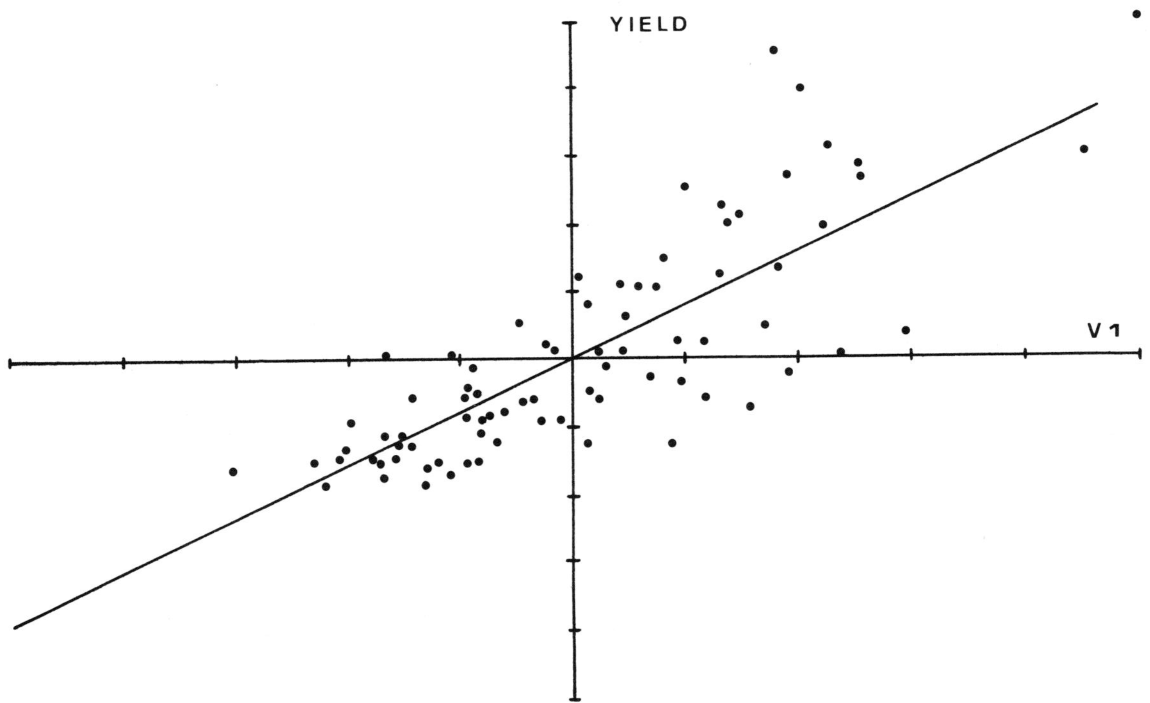

Fig. 3 Principal component analysis for the F_2s.

TABLE 3

PRINCIPAL COMPONENT ANALYSIS. F_2 GENERATIONS RELATIVE ORDER OF IMPORTANCE OF THE MAIN CHARACTERS INFLUENCING YIELD (DIRECTION IS ALWAYS POSITIVE)

Character	F_2									
	172 x 171		172 x 170			171 x 170	171 x 166	166 x 172	165 x 172	164 x 165
	A	B	A	B	C					
Seeds/pod	-	-	-	-	-	-	-	2	-	5
Pods/node	2	3	2	4	2	3	3	1	3	4
Seeds/plant	1	1	1	3	1	1	1	-	2	2
Pods/plant	4	2	3	2	4	2	2	3	1	1
Seed weight	-	-	-	-	-	-	-	-	-	-
Seeds/ovule	-	5	5	-	5	5	5	5	-	-
Ovules/ovary	-	-	-	-	-	-	-	-	-	-
Flowers/node	-	-	-	5	-	-	-	-	-	-
Pods/flower	3	4	4	1	3	4	4	4	4	3
Days to flower	-	-	-	-	-	-	-	-	-	-
Orob./plant	-	-	-	-	-	-	-	-	5	-
Number of plants	80	60	77	75	60	94	67	79	64	70
% variation explained by the first vector	31	31	34	39	40	36	26	30	30	41
Correlation with yield	.82	.86	.88	-	-	.83	.77	.78	-	-

(d) Final conclusions

The results for F_2 plants are clearly different from those obtained with pure lines and F_1s and are more appropriate to the practical selection methods used by plant breeders. It should be recognised that the results could have been different had different parental lines for the F_2s been chosen. However, the results in Figure 3 also suggest that, at least in our conditions, a similar result could equally well have been obtained.

It is interesting to note that some characteristics, such as seed weight, seeds/pod etc., were not totally unimportant. It is well known that at least seed size is an important yield component. This fact is a good example of the limitations of this kind of study; all crosses which included a *paucijuga* line as a parent (171 and 172) produced small seeds because small seed size is dominant; the F_2s were skewed towards the *paucijuga* type in Figure 3, and the large seed size was not recovered even in the case of 165 *(major)* x 172 *(paucijuga)* (165 does not have a big seed). On the other hand, 164 and 165 have similar seed sizes. Thus, in our study seed size did not show much variation, resulting in other factors influencing yield.

It is possible to explain the behaviour of other characters in a similar way. Thus, seeds/pod did not show great variation among parents either, and it is likely that if lines with 7 - 9 seeds/pod had been used, this would have been reflected in the results.

To conclude, the following factors emerge:

(1) The correlation between yield and the values given by a linear combination of characteristics based on the eigen-vectors obtained after a factorial analysis, is a valuable method of studying yield components as determinants of yield.

(2) Factors influencing yield differ depending on the genetic material used. The differences in genetic architecture responsible include: the existence of 'blocks' of characters (mainly if cultivars or pure lines are studied), the existence of a character showing insufficient variation, the peculiarities of the hybrids (mainly their self-fertility) etc.

(3) If the materials include segregating generations and/or lines derived from these, it is necessary to interpret the results bearing in mind the genetics of the characters included in the study.

(4) Emerging as a concrete result of the application of this method to a set of F_2 plants including parents from the four botanical groups was the finding that pods per plant and pods per flower were the most practical characters to be used in selection.

REFERENCES

Cubero, J.I., 1973. Evoultionary trends in *Vicia faba* L. Theor. Appl. Genetics, 43, 59-65.

Kendall, M.G., 1972. A course in multivariate analysis. Allan Stuart, London.

Martín, A., 1976. Genética de los componentes del rendimiento en *Vicia faba*. Ph. D. Thesis, Escuela Técnica Superior de Ingenieros Agrónomos. Córdoba.

Martín, A. and Cubero, J.I., 1978. Inheritance of Quantitative Characters in *Vicia faba*, in 'Some current research on *Vicia faba* in Western Europe. Ed. by the European Committee. pp. 383-395.

DISCUSSION

M. Frauen *(FRG)*

I have a more general question about the dominance of seed size. Were these paucijuga crosses made with large seeded types or is it a general inheritance characteristic that the small seed size is dominant?

J.I. Cubero *(Spain)*

The type I am referring to is the inheritance of the size in the paucijuga lines. We are at present selecting F_3's and F_4's from crosses formed between our smallest paucijuga and the variety with the largest seed. The cross was fertile and we are now selecting for yield. The maximum seed size is similar to that of equina type but very few F_3 or F_4 lines are so large. Most of them are about the size of minor or even less.

M. Frauen

Have you carried out any reciprocal crosses, or are these crosses in one direction?

J.I. Cubero

In F_1 there are reciprocal differences but it is difficult to say the same for F_2, F_3 or even F_4.

M. Frauen

So in F_2 and F_3 there are no reciprocal differences?

J.I. Cubero

I cannot say. If they are of significance, they are much less. But with F_1 you get reciprocal crosses, yes.

R. Thompson *(UK)*

As regards the number of seeds per pod, it seems to be an extremely stable character in relation to environment.

J.I. Cubero

To a certain extent. There are cultivars which were selected long ago - even centuries ago - for long pods; there you get 7 - 8 seeds per pod. There is someone in southern Spain who selected for extra large pods. In one generation he got ten seeds per pod, but the following generation was completely infertile. It seems that there is a limit. My first studies with *Vicia faba* were with Aquadulce in Spain. The differences between Aquadulce strains were immense but they are stable for the number of seeds per pod.

G.G. Rowland *(Canada)*

When you say that you are selecting for yield in the F_3, are you selecting the yield of individually spaced plants or are you selecting from rows?

J.I. Cubero

The system has been to make the cross and to cultivate F_1 and F_2 without selection. Basically it is a genealogical system, the best plant being within the best F_3.

G.G. Rowland

And are these spaced plants?

J.I. Cubero

They are wide spaced in F_2 and also in F_3 but from now on different families will be compared at more normal commercial densities.

D.A. Lawes *(UK)*

Can we commence this period with questions which were cut off on any of the papers we have had so far today?

D.A. Bond *(UK)*

I have two comments on Dr. Austin's paper. The topless material which he described was a population which was in the course of selection. Selections by other breeders based on the Svalof mutant may behave quite differently to the topless material which Dr. Austin described.

The second point is that perhaps it would be fair to say that in some other yield trials the yield of the topless has only been 70 - 80% of Maris Bead.

Could he say whether the topless is not an adequate model, in that I understood that photosynthesis of the pods is not very great. We are putting them up on the top of the plant where they are receiving all the light. Will the uppermost expanded leaf take over the job that we had hoped the pods would do had they been photosynthetically efficient? Did your comments about the proportion of carbon assimilated by various organs refer only to the 2nd August sampling date? Is the leaf area duration longer in the uppermost leaf of the topless, compared with the leaves of the Maris Bead? It seems to be a bigger and thicker leaf and one that will remain on the plant longer. Is it particularly efficient and can it contribute as much as the sequentially expanding leaves of Maris Bead?

R.B. Austin *(UK)*

There are quite a number of points there. I should have mentioned the point that it is a population of topless that we used. In relation to the question of the labelling occasions, this was done on four occasions, three of which were during the pod filling phase of which I gave data for one only.

Dr. Bond also asked me to speculate on the usefulness of the topless model. One of the things which struck me was that, although it produced a great many unproductive side-branches, the total dry matter it produced and the yield of straw it produced was, in fact, greater than that of Bead. We were not working with isogenic lines and therefore we cannot be sure that the effects which we measured were the effects of a particular gene. However, it suggests to me that it is not inherently less productive, or less vigorous, and that if one could channel the dry matter that is produced in early growth into the production of fertile branches, this ought to offer the opportunity of achieving at least an equal yield potential. I would have thought that, providing these branches were only formed low down on the plant, then this should be a more stable model than the indeterminate plants which we have at the moment.

Another point I would like to mention concerns the rates of photosynthesis per unit weight of pod. We measured this from their specific activity, and the ^{14}C per gram of dry matter in the pod and pods in topless were photosynthesising at a faster rate than those of Maris Bead. This would only be detrimental if these pods were, by their positions, shading leaves at a lower level on the plant and reducing the contribution of those shaded leaves. I think it is rather unlikely that this would occur in the topless model because, as Dr. Bond has pointed out, the largest leaf on the top of the plant makes the largest single contribution to the photosynthesis of the canopy and it is at, if not above, the level of the pod. I must also stress that I would not like to make any definite pronouncement from one experiment, in one year, with one population of topless plants.

R. Thompson *(UK)*

The TI (Topless) which I reported on earlier was grown in

the same year as Dr. Austin's topless and was obtained from Dr. Bond. I am interested to know whether it was the same TI as Dr. Austin used. Also the total dry matter yield which I obtained based on growth analysis, was lower than that from Cambridge. Was it the same TI?

D.A. Bond

It was the same population, yes.

R. Thompson

Assimilation chamber measurements on mature pods of beans at SHRI showed that they are not net photosynthesisers. Their photosynthesis is fully accounted for by respiration. Therefore, although Dr. Austin obtained apparent photosynthesis in pods of Maris Bead, as he points out, the pods may have been at compensation point, they are rather low in the canopy, and there may well not have been net photosynthesis. Even those at the top of the canopy may not achieve a net contribution to canopy photosynthesis even though they may assimilate more carbon because of their favourable position in relation to incoming radiation. I am not sure that I would agree with him on the question of whether it matters or not that the pod is above and thus shading what is the most efficient photosynthesising organ - the leaf.

The population density used by Dr. Austin was 30 plants/m^2 which is quite low and this would be expected to give a leaf area index which may allow sufficient light to penetrate to the lower pods and enable photosynthesis, which would not be possible in a more normal canopy.

R.B. Austin

I also observed the difference that Dr. Thompson got between the dry matter production of Topless and Maris Bead and that it was not the same as that which we obtained. I was going to ask him to explain it!

On the question of densities, I am not an expert in the agronomy of field beans but we were advised that the density we achieved was the normal density. I would ask Dr. Bond if he would confirm that.

D.A. Bond

We tried to obtain the normal density which is used in our area. However, I would agree that it should have been 40 instead of 30 plants/m^2. I wonder if Dr. Sjödin could comment on the effect of density on the relative yield of topless and normal beans. Did he not find that high density was favourable to this comparison, and that 40 plants/m^2 was not enough to get the two varieties yielding equal amounts of grain?

J. Sjödin *(Sweden)*

When we carried out an experiment some years ago, the yield was increased markedly with an increase in plant density. In Sweden 60 plants/m^2 is the usual density but with a 100 plants/m^2 a better yield was obtained. That has practical implications of course; how much can be sown from an economical point of view? There was another factor, the number of side branches was drastically reduced with an increase in density. We also found that the upper leaves were thicker than the others and the same effect was achieved with decapitated plants.

We have included the topless character in many different genotypes and although considerable differences in performance were found, so far we have been unable to reach a satisfactory level of yield. The best combinations presently give about 80% to 90% of the yield from standard varieties.

I would like to ask Dr. Frauen what happened to the less drastic topless types he found some years ago.

M. Frauen *(FRG)*

Firstly I would like to give some information about my crossing programmes with the Svalöf topless types. This was not a crossing programme with known parents but rather a polycross. I worked with natural outcrossing, plants of the mutants being spread throughout the bean nurseries. After four years some lines are improvements on the original topless types, branching simultaneously with two or three branches. The number of flowering nodes needs to be improved as well as reducing flower distortion.

Two years ago I found a similar type which I called 'semi-determinate' and which had a determinate flower and stopped extension growth, but normally did not produce a pod on the last flowering node. I think we saw similar types in Dr. Bond's nurseries last year in Cambridge. It was a Spanish type and I think it came from Professor Cubero. But the determinate habit was less marked and normally there is no distortion of the inflorescence. I think this type may be more promising, especially in relation to the problems with the flowers. We have great problems with flower development and performance in the topless plants derived from the Svalöv mutant.

D.A. Lawes

Can we open up to a more general discussion now. Would anyone care to voice a general opinion on any of the physiological work that has been discussed today?

R.B. Austin

Could I ask breeders, collectively, whether they could give us a fairly clear statement on the genetic control of seed size and also pods per plant?

D.A. Lawes

Perhaps Dr. Bond would like to comment on this?

D.A. Bond

I would like to pass it on. Dr. de Vries has done the heritability of seed size and components of yield.

A. Ph. de Vries *(Netherlands)*

I cannot comment on the inheritance - whether it is dominant or recessive. But as far as heritability is concerned I think seed size has as high a heritability as number of pods per plant. Therefore, selection on seed size is more successful than selection on number of pods per plant.

D.A. Lawes

We are trying to establish the relative importance of the different types. I believe Professor Cubero would like to come back on this.

J.J. Cubero *(Spain)*

The heritability depends on the parents. It is the same when speaking about dominance or recessiveness - we have to be sure about the parents. In *paucijuga* progeny we are selecting for lines between crosses of large and small seeds and in some the value for this character is lower than that for the parental average. It is true that *paucijuga* is a very special source of genes and I cannot say if the smaller size is dominant in all cases. There could be other genes for a large size that are dominant over small seeds.

We have fewer studies on number of pods per plant. A large number appears to be dominant and there is heterosis in *Vicia faba;* the F_1 has more seeds per plant and pods per plant than the parents.

D.A. Lawes

In the UK, we find that the equina type seed size gives the higher yield and reduction in seed size leads to loss in yield.

D.A. Bond

Leaving aside the *paucijuga* types, it is much easier to get a higher yield from the larger seeded types.

M. Frauen

I have a question for Professor Cubero about the crosses with *paucijuga*. One of the main points raised this morning was the problem arising when *Vicia faba* is grown in ideal environments, where there is a more or less uncontrolled growth; the harvest index goes down and the yield of seeds is generally not any higher than in less ideal environments. *Paucijuga* types are normally of reduced vegetative growth and so is Professor Cubero selecting for types with reduced vegetative growth?

J.J. Cubero

Paucijuga types have reduced leaves. However, when you cross the *paucijuga* lines with any other cultivar, then most of the genes affecting the reduced growth in the *paucijuga* lines are recessive. These *paucijuga* characters are completely lost in F_2 or F_3 generations.

D.A. Lawes

Thank you all for your contributions.

SESSION 2

PLANT MODELS

Chairman: R. Thompson

THE ORIGIN AND DEVELOPMENT OF A PROGRAMME TO BREED LEAFLESS DRIED PEAS

B. Snoad
Department of Applied Genetics, John Innes Institute, Colney Lane, Norwich, UK.

ABSTRACT

At the John Innes Institute the increased standing ability of the so-called leafless pea is being exploited in the development of a dried pea crop suitable for existing canning and packeting markets as well as for reducing some of the EEC's over-dependence upon imported soya.

Leafless peas have their origin in the use of two naturally-occurring mutant forms one of which, with leaflets converted to tendrils, was found in Finland in 1953 and the other, with exceptionally small stipules, in England in 1923.

The only leafless dried pea variety so far registered on the National list, Filby, has in extensive trials in the UK proved to have yields ranging from 1.7 - 6.2 tonnes/ha with an average of 3.4, improved standing ability and no more than 2% stained seed.

The breeding programme is based upon a combination of pedigree breeding and single seed descent. Selection favours early flowering, one or two flowers per node with a target of some 20 - 30 seeds per plant, according to size, which at appropriate plant densities, is all that is required for yields in the region of 5 tonnes/ha. Stiffer stems, non-shattering pods and disease resistance are also sought. Since the protein content of the seed is greatly influenced by environment, selection for protein content is not yet considered to be feasible.

A number of areas for more research effort including protein content, responses to environment, incidence of pests and diseases, control of weeds, the importance of mycorrhizal relationships and the specific uses for pea seed protein and starch are mentioned briefly.

INTRODUCTION

There is no doubt that within the EEC we should make every effort to develop a home-grown, relatively high protein grain crop which can go at least some way towards reducing our over-dependence upon imported soya. Among the established legume crops we have the limited choice of field beans, peas, *Phaseolus* beans and to a certain extent, soya, all of which have their problems, but among the less well established crops such as lupin and soya in Northern Europe, the problems are even greater.

Any opportunity that presents itself and introduces the possibility of making significant improvements to our established crops should therefore be seized, tested and exploited with the minimum of delay. Such an opportunity now exists with the synthesis of the so-called 'leafless peas', which, in contrast to the intuitive predictions that have been made, are already proving to be capable of giving crop yields as good as those of conventional peas with the added advantages of easier harvesting and improved seed quality.

The purpose of this paper is to describe briefly the origin of this novel phenotype and to outline the aims and objectives of a recently introduced breeding programme devoted to its development as a dried pea crop in the UK.

THE ORIGIN OF LEAFLESS PEAS

Leafless peas differ from conventional peas in being recessive at the af locus (leaflets entirely replaced by tendrils; Plate I) and the st locus (stipules much reduced in size; Plate I). Naturally-occurring mutants of these two kinds have been known for many years; af in Finland (Kujala, 1953) and st in England (Pellow and Sverdrup, 1923) but although the two genes have been combined to develop new genetic stocks they seem not to have been exploited in the breeding of new pea varieties.

Plate I: A. Semi-leafless pea (afaf.STST); B. Leafless pea (afaf.stst); C. Leafless pea Filby showing improved standing ability of crop prior to harvest.

The use of the <u>af</u> gene alone, to develop what is known in England as the 'semi-leafless' pea (Snoad and Gent, 1974, Plate IA) did take place in Russia to develop a variety which they called the 'moustache pea' (Solovev, 1958). The combination afaf.stst, the 'leafless pea' (Snoad and Gent, 1974; Plate IB) seems consistently to have been ignored probably because individual plants of this genotype when grown in well-spaced conditions against supports always yield less seed than conventional or semi-leafless peas. In a closely sown field crop, however, whilst the yield of individual conventional pea plants falls dramatically, that of the leafless peas is less affected by competition so that the yield of seed per unit area can be similar in both types of plant.

It is the greatly improved standing ability of the crop, due to the inter-twining of the many tendrils with each other and with the stems, that leads to our interest in this new phenotype (Plate IC). Peas characteristically lodge and are therefore particularly difficult to machine harvest. Because of the lodging much of the yield can be lost and the quality of the seed is often poor due to staining and pathogen attack arising from the damp soil conditions. We therefore now have a new plant type which should not only make harvesting easier but also provide seed of a better quality.

A programme to breed leafless peas has now been in existence at the John Innes Institute for just over four years. Its aim is primarily to reduce our imports from N. America of Alaska type peas which are used in large quantities in the canning industry and also to improve the harvestability, yield and quality of the other types of dried peas which are used for packeting, in soups and in convenience foods. At the same time it is realised that the leafless pea is one of the candidates for the substitute for soya which is needed within the EEC. The foundation for this programme is the John Innes Pea Collection which now numbers over 1 300 distinct accessions gathered from around the world.

Although this breeding programme is relatively new it is based upon an extensive research programme in which a wide range of novel pea phenotypes was developed between 1970 and 1976 (see Snoad et al. in Annual Reports of the John Innes Institute nos. 61 - 67) and tested for their agricultural suitability. One of these types, a leafless pea, proved in a series of field trials carried out entirely by other organisations to have sufficient potential for it to be considered as a variety in its own right; this is the cv. Filby. Information is steadily accumulating about the performance of Filby in a wide range of sites and many differences continue to be demonstrated between leafless and conventional peas in their husbandry requirements. There is a lot of information too which is helpful in shaping the breeding programme itself; what crosses to make, what selections are needed and what environments are appropriate for both selection and yield trials.

THE PERFORMANCE OF THE LEAFLESS PEA FILBY

Before discussing the breeding programme it is appropriate to provide a summary of some of the information which has accumulated about the prototype leafless pea, Filby.

Filby has been grown over a period of some four years in trials carried out over a wide range of agricultural environments (Table 1). Its yield in trials where Vedette has been included as a control is indistinguishable from that of Vedette and possibly slightly less variable. These figures have to be considered against the average UK dried pea yield for the last five years for which data are available and which is 3.13 tonnes per hectare so that even taken over all countries and years the yield of Filby is still acceptable. In looking at these figures it is also necessary to know that the plant densities achieved in the majority of these trials were considerably lower than those that would now be recommended. In other words, better average yields than these are to be expected from crops grown at higher densities.

TABLE 1

SEED YIELDS OF FILBY, AND OF VEDETTE IN ELEVEN TRIALS, OBTAINED BETWEEN 1975 AND 1979

Name (and trial country)	Number trials	Dry seed yield (tonnes/ha)	
		Mean ± SD	Range
Filby and	11	3.6 ± 0.54	2.8 - 4.8
Vedette (UK)	11	3.4 ± 0.87	2.0 - 5.0
Filby (UK)	42	3.4 ± 0.94	1.7 - 6.2
Filby (Canada and Finland)	6	2.8 ± 1.43	1.2 - 5.4

The amount of seed staining reported has consistently been low in Filby, averaging between 1% and 3%.

The standing ability of cv. Filby has generally been better than that of conventional pea crops. There have been however a small number of instances, and particularly involving three sites, in which standing although protracted has eventually failed by harvest time. There are a number of factors any of which might have singly or collectively been associated with these crop collapses including very fertile soil, high moisture content of the soil, too low a crop density, inefficient weed control, sowing too late and delayed harvesting. These are topics which obviously need careful investigation in the near future.

Weed control is more demanding with Filby but usually a pre-emergence herbicide has been the only treatment required on average soil types.

THE LEAFLESS PEA BREEDING PROGRAMME

Leafless peas are unique so that it is impossible initially to specify either an ideotype or a selection enviroment and the programme cannot therefore be an extension of a programme to breed conventional peas. Maximum variation is obviously of prime importance and this is being obtained by means of pedigree breeding rather than back-crossing in which it would be impossible to specify the ideal recurrent parents. Within the last two years some use has also been made of single seed descent because initial selection is then primarily centred upon a small crop of peas at about the F8 stage which is likely to be more meaningful than single plant selection at every stage of development.

An important point to be borne in mind when breeding peas is that the high potential yields of well-spaced individual plants will never be maintained in a field crop. When deciding how many yield components each plant should carry therefore, it is essential to be aware that at a density of about 100 plants/m^2 each plant needs to carry only 20 - 30 seeds in order to provide crop yields well in excess of those currently attained (Table 2). A realistic number of ovules per plant which, when developing into seeds, does not result in excessive competition for nutrients in quantity, quality or time is all that is required. It is too early to make precise recommendations for leafless peas but already there is good evidence for developing plants with no more than two flowers per node (Snoad and Arthur, 1974). The number of ovules per pod should probably be as high as possible but this is certain to be influenced by seed size itself and other factors such as the competition between pods at a node or pods at successive nodes.

There are certain specific requirements for the vegetative framework of the leafless pea plant. When grown at an optimum density any basal branching will soon be suppressed and it therefore seems logical always to select against this character. The type of stem is also important and experience has shown that thick stems which characteristically possess large cavities tend to be weak and are likely to lead to crop collapse. Thin stems can be developed genetically but there is also a strong environmental component involved as a result of interplant competition in the sward. Thin but stronger-stemmed plants grown at relatively high density are more likely to be able to support the relatively small number of seeds that each develop. Since there seems to be some variation in stem strength *per se* this is also being introduced into the programme for evaluation.

As with conventional peas there is extensive variation for both 'leaf' size and stipule size. At the moment there is little to indicate the ideal phenotype and so as wide a range of types as possible is being synthesised for experimentation.

TABLE 2

THE NUMBER (AND TOTAL WEIGHT) OF PEA SEED NEEDED PER PLANT, ACCORDING TO SEED SIZE AND CROP DENSITY, IN ORDER TO ACHIEVE A YIELD OF 5 TONNES/HA

	Weight of 1 000 seeds (g)						
Density	100	150	200	250	300	350	Seed weight per plant (g)
50/m^2	100	66	50	40	33	29	(10)
100/m^2	50	33	25	20	17	14	(5)
150/m^2	33	22	17	13	11	10	(3.3)
200/m^2	25	17	13	10	8	7	(2.5)

Pod splitting, or shattering, when the crop is ready for harvesting could be a problem leading to seed losses. Attempts are being made to overcome this in two ways; first by developing so-called sugar peas in which no inner sclerenchymatous layer develops and secondly by using a modification to the funicle as a result of which the dry seed remains more firmly attached to the inside of the pod even when it has opened.

Resistance to disease is important and the programme includes routine attempts to breed for resistance to *Fusarium oxysporum*, *Ascochyta pisi* and *Peronospora viciae*.

The type of seed required varies from industry to industry. There is nothing to suggest that there are limitations to developing leafless pea seed of any size, colour or shape but there may be a limitation on seed size imposed by the plant itself since large seeds tend to be developed upon large plants which in turn tend to have poorer standing ability as crop plants. It may of course be possible to strengthen these larger stems but even so, as my colleague Dr. Hedley will show, maximum yields from leafless peas seem more likely to be derived from high density crops with small to medium seeds rather than large ones.

The whole subject of the leafless pea and its environment for selection, trial and growth is a fascinating and complex one. With a novel form of plant such as this, the obvious way to begin seemed to be by growing it in the same way as small conventional pea plants. This however has proved to be somewhat inappropriate but I will leave it to Dr. Hedley to describe the type of leafless plant and the form of environment which the results of his research programme suggest are likely to provide dried pea yields as good as, or even better than those we currently obtain. Fortunately there seems to be complete compatibility between the precise experimental findings of the crop physiological approach and the more empirical ones of the plant breeder.

Finally there is the all-important subject of pea seed protein. Pea seeds have the advantages of very low levels of trypsin inhibitor, phyto-haemagglutinin, lipoxygenase and urease activity when compared with soya and they are very similar to those of soya in methionine and cystine content but contain approximately 30% more lysine.

The biggest difference between peas and soya is in protein content; that of peas not exceeding 32% (Matthews et al., 1975; Jermyn and Slinkard, 1977) and it is most unlikely that pea protein content will ever be raised to the 40% level of soya. As far as Europe is concerned the big advantage of peas lies in their grain yield which results in the yield of protein per unit area being equivalent to that of a N. American crop of soya. Any improvements in pea seed yield will thus automatically improve protein yield and this in the main and immediate theme of the John Innes programme, for reasons which will become apparent below.

As with the leafless pea and its 'growing' environment there is an equally fascinating and productive area for research into the relationship between seed protein content and environment. It is well established that the environment in which a pea plant is grown has a strong influence upon the protein content of the seed but very few attempts have been made to obtain the fundamental type of information which would help the breeder in his search for high grain yield, high protein content and, above all, relatively stable response of both to environment. It is very important to determine how much variation there is from seed to seed, from pod to pod, from plant to plant and from environment to environment since without this type of information, comparisons and correlations can have little meaning.

Some preliminary studies of this kind have been undertaken at the John Innes Institute (Matthews et al., 1975) and the protein content of seed samples from 255 pea genotypes growing in one environment estimated as ranging from 10 - 32% (Figure 1A). These values are based upon the measurement of α-amino nitrogen using the ninhydrin reaction, a technique in which the error is no more than ± 5%. In addition, considerable variation in protein content was observed between samples obtained from different parts of the same plant and from different plants of the same genotype (Figure 1B and C and Table 3).

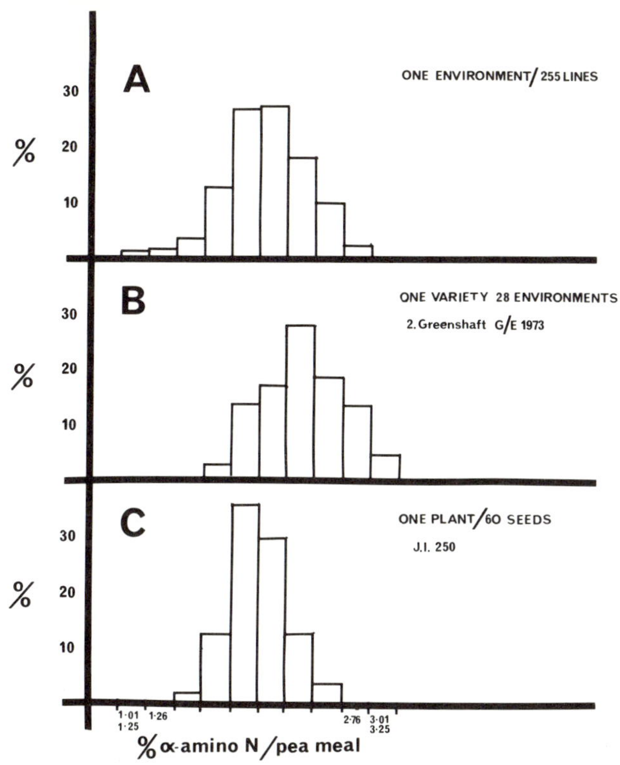

Fig. 1. Percentage α-amino nitrogen in the meal of pea seeds. A. 225 genotypes grown in one environment. B. One genotype grown in 28 environments. C. Sixty individual seeds from one plant. (Reproduced by permission of P. Matthews).

TABLE 3

VARIATION IN PROTEIN CONTENT ESTIMATED IN PEA SEED MEAL SAMPLES (P. MATTHEWS, PERSONAL COMMUNICATION)

Between individual pods of one plant	± 8%
Between individual seeds of one plant	± 30%
Between individual plants of one genotype:	
a) in one environment	± 20%
b) in a range of environments	± 40%

Since environment has such a marked effect upon seed protein content the plant breeder cannot select for it in the same way as he does for other characters. When populations of relatively homogeneous material eventually become available, however, the protein could then be estimated from crops grown on a number of sites in order to determine mean values and ranges.

Evidence such as this serves to emphasise the importance at this stage of breeding primarily for plant performance and for grain yield but at the same time there are other areas of research which are particularly important if we are to make even more progress.

Variation in pea seed protein content and the relationships between genotypes and their environments should be examined closely in the hope of identifying and introducing a degree of genetically controlled stability into the system.

The rapidly accumulating variation which is being developed in the leafless pea breeding programme should be exploited in order to extend the number of crop physiological studies being undertaken and thus speed up the logical progression towards a more agronomic approach.

The response of leafless peas to nutrients, to drought and to water logging could be different and should be examined in more detail.

The incidence of pests and diseases in leafless pea crops has not been determined to any extent and certainly not in the wide range of environments such as we have within the EEC.

The control of weeds requires more careful examination since ground cover during early sward development is much poorer than in a crop of conventional peas.

The importance of mycorrhiza in the uptake of phosphates by peas and in influencing the efficiency of nitrogen fixation by Rhizobium has not been sufficiently recognised.

Finally, there is the question of the end product itself; the seed. We need to know much more about the range of animals for which it is likely to be suitable, the specific needs for changes in protein quality and about the requirements for protein isolation, spinning, texturisation, etc. Since about 75% of the seed is composed of starch it would be useful also to consider the uses to which it could be put together with any requirements for changes in its quality.

REFERENCES

Jermyn, W.A. and Slinkard, A.E., 1977. Variability of percent protein and its relationship to seed yield and seed shape in peas. Legume Research 1 (1), 33-37.

Kujala, V., 1953. Felderbse, bei welcher die ganze Blattspreite in Ranken umgewandelt ist. Arch. Soc. Zool. Bot. Fen. 8, 44-45.

Matthews, P., Dow, K.P. and Rumary, C., 1975. Protein quantity and available methionine in peas. Ann. Rept. John Innes Institute, no. 66, 26-32.

Pellew, C. and Sverdrup, A., 1923. New Observations on the genetics of peas. Genetics 13, 125-131.

Snoad, B. et al. (1970 - 1976). Ann. Repts. John Innes Institute, nos. 61-67.

Snoad, B. and Arthur, A.E., 1974. Genotype - environment interactions in peas. Theor appl. Genetics 44, 222-231.

Snoad, B. and Gent, G.P., 1974. Practical assessment of new pea phenotypes. Ann. Rept. John Innes Institute, no. 65, 22-23.

Solovev, V.K., 1958. 'New forms of vegetable peas'. Agrobiologia 5, 124-126.

DETERMINING IDEOTYPES FOR THE 'LEAFLESS' PEA CROP

C.L. Hedley
John Innes Institute, Colney Lane,
Norwich, UK.

ABSTRACT

Significant differences have been found between the responses of leafed and 'leafless' pea plants when grown at different planting densities. The biological and economic yield per unit area of the 'leafless' phenotype decreased at low planting densities because the plants were unable to compensate fully to planting densities below 100 plants/m^2. It is essential, therefore, that 'leafless' plants are grown at high population densities; so the ideal 'leafless' crop plant will be tolerant of inter-plant competition and therefore, by definition, be a weak competitor. The 'leafless' ideotype which will tolerate these conditions is defined as having small seeds, to reduce plant growth-rate; non-branching, since branches add little to yield at high density; early flowering, so that partitioning of assimilates is initiated when within-plant competition is low; indeterminate habit, to enable the crop to attain maximum biomass per unit area; have single fruits per node, to reduce competition between nodes; have small seeds with a low relative growth-rate to reduce competition within the pod and the fruit should be efficient at fixing and recycling carbon to maximise the efficiency with which photoassimilate is utilised at each node.

INTRODUCTION

For most crops there is little evidence that physiological information has directly assisted plant breeders. The role of the physiologist has invariably been retrospective, giving the breeder physiological explanations of attributes which have been inadvertently incorporated into improved varieties. The introduction of the 'leafless' phenotype into the dried pea breeding programme, however, has presented the breeder with unique problems which may require physiological information about the plant and about the crop before suitable genotypes can be selected.

The problem of how to predict the effects of the 'leafless' phenotype on characters which can only be derived from conventional peas, has been partially overcome by incorporating the genes for 'leaflessness' (st and af) in as wide a range of genetic backgrounds as possible. The problem still remains, however, of how to select, from large numbers of single plant segregants, plants which will produce good yields when grown in a crop environment. There is no established 'leafless' pea crop and very little information about the 'leafless' phenotype on which the breeder can base his ideas. Neither is there any reason to suppose that plants with the 'leafless' phenotype will behave in a similar way to the established leafed varieties. One method of alleviating this problem is for the physiologist to formulate, in collaboration with the breeder, plant 'ideotypes' which will make good crop plants. This is a similar approach to that used by Donald (1968) for wheat, although a great deal of information existed about the behaviour of wheat as a crop plant.

BEHAVIOUR OF THE 'LEAFLESS' PHENOTYPE WHEN GROWN AS A CROP

Comparisons between several leafed and 'leafless' genotypes have revealed significant differences between the two phenotypes when grown as crops (Hedley and Ambrose, in preparation). Although individual plants of both phenotypes increased in total biomass with increased available space (decreased density) the leafed plants increased more rapidly and, at 16 plants/m^2, weighed on average twice as much as comparable 'leafless' types (Figure 1a). At planting densities in excess of 100 plants/m^2, however, the total biomass of individual 'leafless' plants exceeded that of the leafed types. The differences between the two phenotypes are more clearly seen when expressed as biological yield per unit area (Figure 1b). The biomass per unit area of the 'leafless' phenotype increased with increased planting density, while the leafed phenotype tended to decrease at higher densities. The 'leafless' biomass per unit area surpassed that of the leafed phenotype at densities greater than 100 plants/m^2 (Figure 1b).

Similar differences between the two phenotypes were found for economic yield (Figures 2a and 2b) except that the yield from 'leafless' plants attained a maximum per unit area between 100 and 400 plants/m^2 and then decreased (Figure 2b). The maximum yield per unit area attained by the 'leafless' phenotype at these high planting densities was equivalent to the maximum yield obtained from the conventional phenotype. It is apparent, therefore, that unlike most crops, including conventional peas, where the planting density giving the optimum biological yield is similar to that for optimum economic yield (Donald, 1963), in 'leafless' peas the density at which these two optima occur differs.

From the observations made on the 'leafless' phenotype grown in swards, two broad requirements of the ideotype can be defined. Firstly, in order to attain an acceptable biological yield per unit area, the plant must be capable of growth at high planting densities. Inter-plant competition must, therefore, be minimised and the plant will then, by definition, be a weak competitor. Secondly, the efficiency with which the plant partitions biomass into economic yield must be maintained at these high planting densities and intra-plant competition therefore minimised.

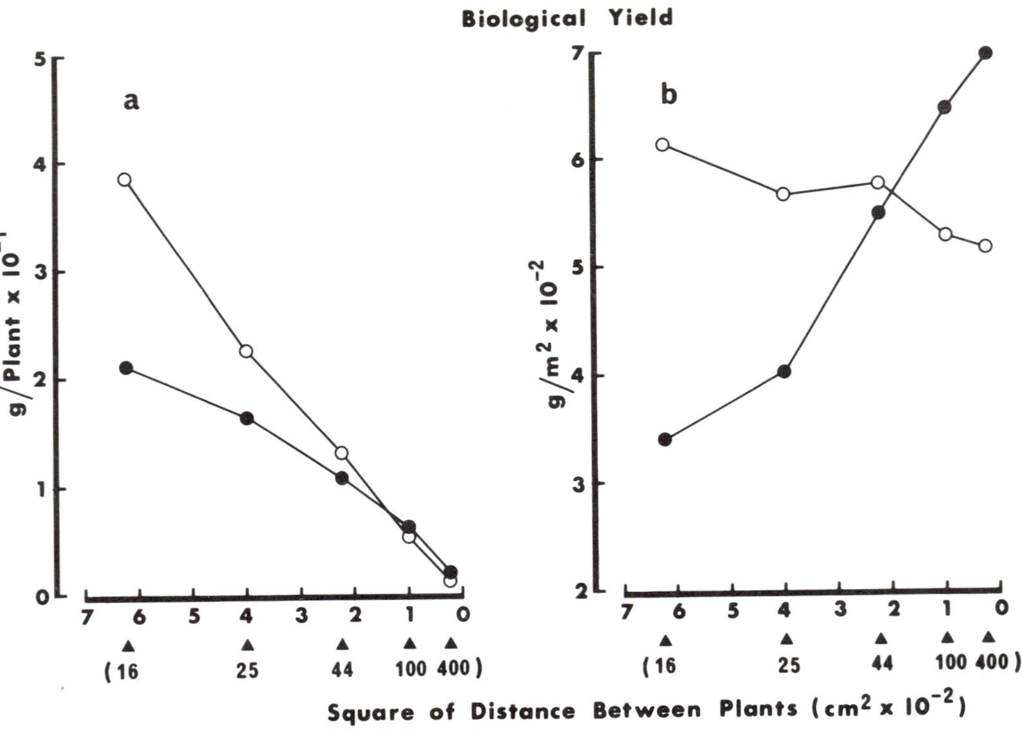

Figs 1a and 1b: Effect of available space on the total biological yield (dry weight) per plant (a) and per unit area (b) of leafed (o——o) and 'leafless' (●——●) pea phenotypes. Numbers in parenthesis are the planting densities/m^2. Each phenotype is represented by the mean of three genotypes.

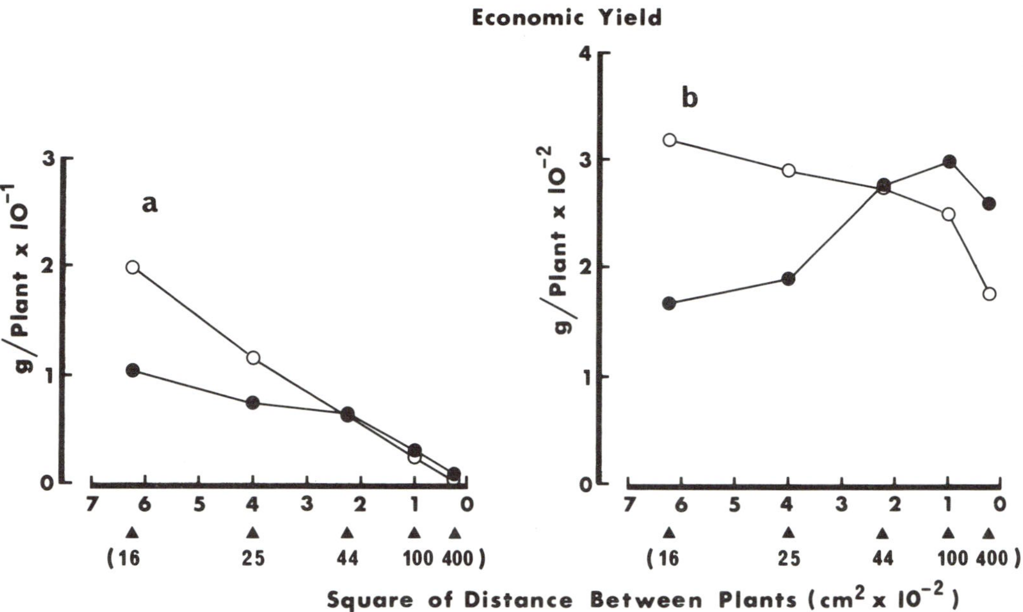

Figs 2a and 2b: Effect of available space on the economic yield (dry weight) per plant (a) and per unit area (b) of leafed (o——o) and 'leafless' (●——●) pea phenotypes. Numbers in parenthesis are the planting densities/m^2. Each phenotype is represented by the mean of three genotypes.

1) Identifying weak competitors

The most important plant characteristic affecting tolerance to inter-plant competition appears to be seed size. Comparisons between 'leafless' genotypes which differed for seed size revealed a strong positive relationship between increasing seed size and an inability to increase yield per unit area at high planting densities (Figure 3a and 3b). The slope for the biological yield response per plant, between high and low planting densities, increased in steepness with increases in seed size (Figure 3a), and this was reflected in a flat response per unit area for the large seeded types and a steep response per unit area for genotypes with small seeds (Figure 3b).

The association of seed size with tolerance to high planting density is symptomatic of the relationship between plant growth rate and plant competition. Over a wide range of leafed and 'leafless' genotypes, within a given environment, there appears to be very little variation for relative growth rate (RGR), as determined from the slope of the natural logarithm of plant dry weight against time (Figure 4). There are, however, large differences in plant size as determined from the position of the regression lines from the x-axis (Figure 4) and these are correlated with seed size. In general, 'leafless' genotypes have lower absolute growth rates than leafed plants; there is a positive relationship, therefore, between seed size and growth rate within each phenotype, but not when the two phenotypes are compared. This is probably because the true relationship exists between the size of the embryonic axis of the seed and plant growth rate and it is likely that the axes from the leafed plants are larger than those from 'leafless' in seeds of comparable sizes.

2) Improving partitioning at high planting densities

(i) Effect of flowering time. Differences in flowering time between genotypes will introduce competition between the reproductive and vegetative parts of the plant at different times in the development of the crop. The crop growth rate

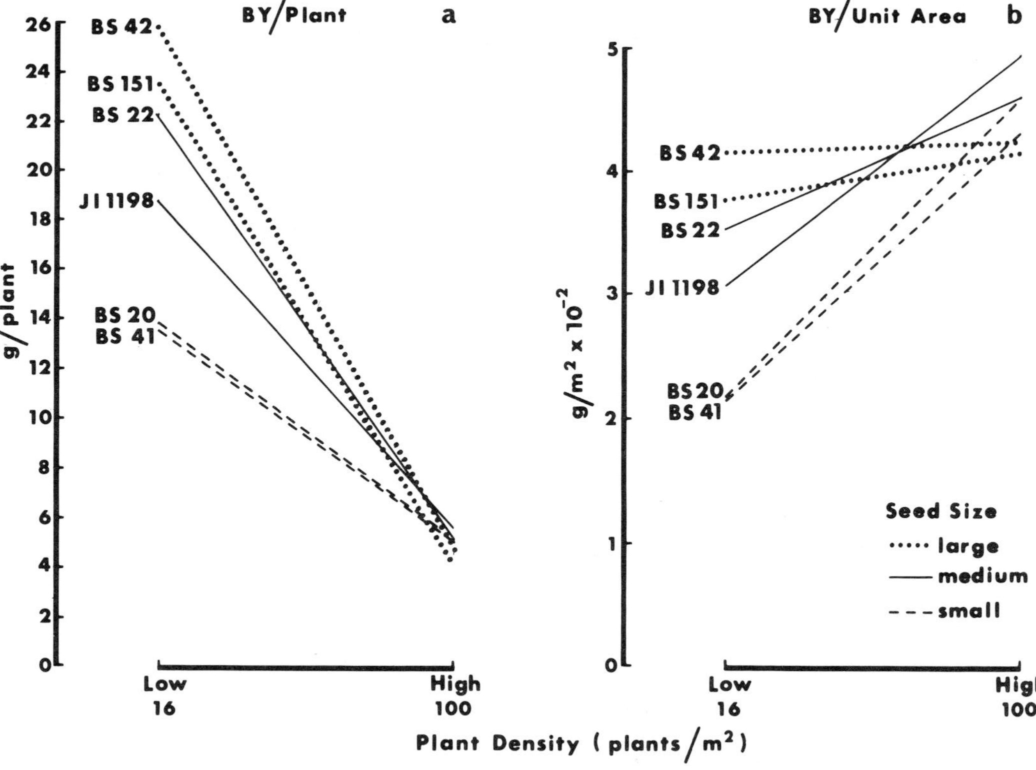

and 3b: Effect on the total biological yield (dry weight) per plant (a) and per unit area (b) at high and low planting densities of 'leafless' genotypes which differ for seed size.

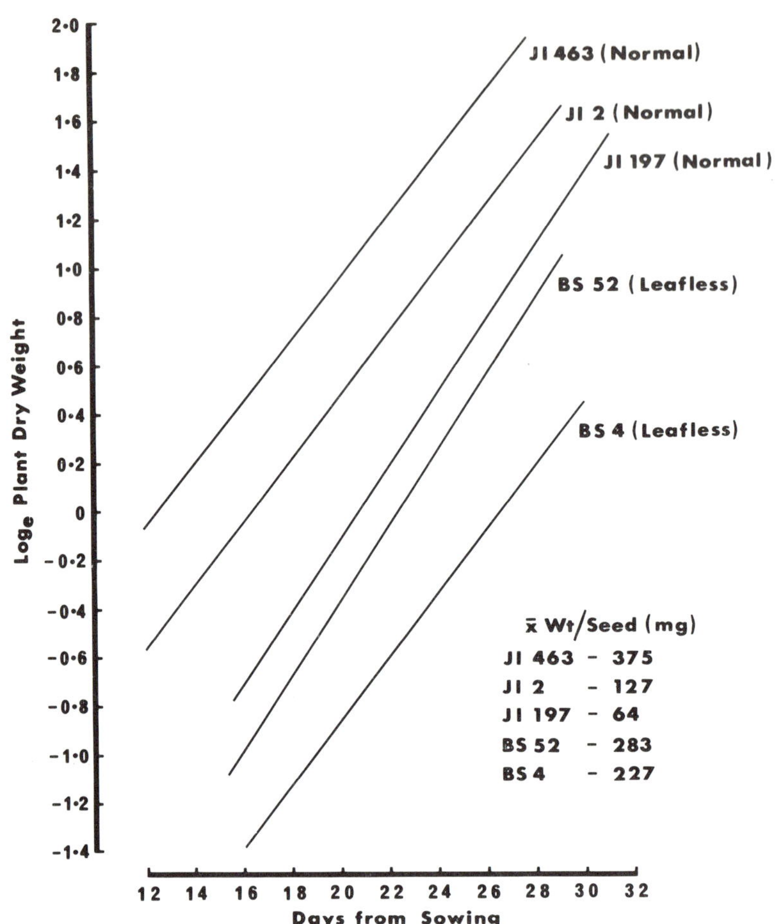

Fig 4: Changes in total biological yield with time for leafed and 'leafless' genotypes which differ for seed size.

of genotypes, which have the same seed size and are sown at the same density, will be similar. If, however, such genotypes differ for flowering time, then the biomass of the crops at specific times may still be similar but the composition of the biomass will differ, as will the effects of intra-plant competition. Accumulation of economic yield in early flowering types will occur over a longer period and will be less subject to the rapid curtailment of crop growth as the season ends. In addition, partitioning in early flowering genotypes is initiated when plant growth rates and competition between plants are low and this will be reflected in a lower level of intra-plant competition. The advantages of early flowering plants, however, will only be made manifest if other physiological constraints are removed, such as premature curtailment of growth due to a determinate habit.

(ii) Effect of yield components. Variation exists for all of the yield components, the final yield being a function of the effect of environment on these components and on the interactions between them. Pea genotypes may differ in flowering time from very early types, flowering at about the seventh node to late flowering types which remain vegetative until twenty or more nodes have been formed. Once flowering has been initiated genotypes may differ in the number of reproductive nodes produced from about four to an almost indefinite number. There appears to be a strong interaction between the degree of reproductive determinancy that a genotype exhibits and the number of flowers formed at each node. Very indeterminate genotypes tend to have one or, at the most, two flowers per node. The flower development at successive nodes is usually separated in time, such that flowers are fully developed at only one or perhaps two nodes at any time. Very determinate genotypes, however, may be multiflowered at each node or the flowers at successive nodes may develop simultaneously, or genotypes may have multiflowered nodes with almost synchronous development.

The asynchronous flowering of the indeterminate habit ensures that competition between fruits developing at each node is kept to a minimum but the development of yield is prolonged. Competition between fruits of genotypes with a determinate habit is more intense and the time over which yield develops is more condensed. As well as the interactions between fruits at a node and between nodes, competition also occurs between individual seeds within each pod. This is most noticeable in comparisons between genotypes which differ for seed size. As final seed size is increased, the number of mature seeds per pod tends to be lower and there is an increased risk of seed abortion. In very large seeded types, very few seeds develop to maturity within each pod and seed abortion is common even amongst seeds which are relatively well developed.

(iii) The effect of the efficiency of the pea fruit. The requirement that 'leafless' peas must be grown in highly competitive environments to attain acceptable yields per unit area make it imperative to select fruit and seed types which are more efficient and more tolerant of such conditions. The pod wall has been shown to fix and recycle carbon within the pea fruit (Lovell and Lovell, 1970; Flinn and Pate, 1970) and the importance of this system to the carbon economy of the fruit has been estimated at 20% (Flinn et al., 1977). Pod walls contain both photosynthetic and dark carbon fixation systems (Hedley et al., 1975; Atkins et al., 1977) and variation for the relationship between these systems exists (Price and Hedley, 1979). It is likely, therefore, that efficient pod types more suited to the 'leafless' canopy can be selected.

Similarly, variation exists for the morphology and physiology of the seed. Unlike the plant, variation for seed RGR can be found in different genotypes (Hedley and Ambrose, in press). It is possible, therefore, to decrease the growth rate and hence the sink demand of the seed by selecting types with a reduced RGR and an increased duration of growth. Alternately, small seeded types could be selected which, relative to larger seeded types with the same RGR, will have lower growth rates and have a lower sink demand when mature.

It may also be possible to improve the efficiency with which the seed utilises assimilate, since the seed has also been shown to have significant levels of the dark carbon fixation enzyme, PEP carboxylase (Hedley et al., 1975). The significance of this system is not known but it can be suggested that the provision of C4 acids, from a recycling system within the seed at a time when protein synthesis is high, may play a significant part in the development of the embryo and may also be important in selecting seeds with a higher storage protein content.

AN IDEOTYPE SUITABLE FOR HIGH POPULATION DENSITIES

The ideotype for 'leafless' peas is to some extent predetermined by the need to grow the plants at high planting densities, but many of the features will apply equally to the leafed crop.

1. The plant will have a reduced growth rate, which ultimately means a small seed size unless variants with a reduced RGR can be found, or unless the relationship between seed size and the size of the embryonic axis can be broken.
2. The plant should be non-branching, since branches do not add significantly to yield at high planting densities (Hedley and Ambrose, in preparation).
3. The plant should be early flowering, so that partitioning into economic yield is initiated when competition is low.
4. Once flowering has been initiated the plant should be indeterminate, to allow the maximum biomass to be produced per unit area and to increase the duration of partitioning into economic yield.
5. Each reproductive node should contain a single pod (or at the most two), to minimise competition between successive reproductive nodes and within specific nodes.
6. The seed should be tolerant of inter-seed competition, to minimise competition within each pod. This will entail selecting seeds which have a low RGR and/or are small.
7. The fruit should be efficient at fixing and recycling carbon, if it can be shown that this recycling system is important for maximising the efficiency with which photoassimilate is utilised at each node.

REFERENCES

Atkins, C.A., Kuo, J., Pate, J.S., Flinn, A.M. and Steele, T.W., 1977. Photosynthetic pod wall of pea (*Pisum sativum* L.). Distribution of carbon fixing enzymes in relation to pod structure. Plant Physiol. 60, 779-786.

Donald, C.M., 1963. Competition among crop and pasture plants. Adv. Agron. 15, 1-118.

Donald, C.M., 1968. The breeding of crop ideotypes. Euphytica, 17, 385-403.

Flinn, A.M., Atkins, C.A. and Pate, J.S., 1977. Significance of photosynthetic and respiratory exchanges in the carbon economy of the developing pea fruit. Plant Physiol. 60, 412-418.

Flinn, A.M. and Pate, J.S., 1970. A quantitative study of carbon transfer from pod and subtending leaf to the seeds of the field pea (*Pisum arvense* L.) J. Expt. Bot., 21, 72-82.

Hedley, C.L. and Ambrose, M.J., 1980. An analysis of seed development in *Pisum sativum* L. Ann. Bot. (in press).

Hedley, C.L., Harvey, D.M. and Keely, R., 1975. Role of PEP-carboxylase during seed development in *Pisum sativum*. Nature, Lond. 258, 352-354.

Lovell, P.H. and Lovell, P.J., 1970. Fixation of CO_2 and export of photosynthate by the carpel in *Pisum sativum*. Physiologia Plantarum, 23, 316-322.

Price, D.N. and Hedley, C.L., 1980. Developmental and varietal comparisons of pod carboxylase levels in *Pisum sativum* L. Ann. Bot. 45, 283-294.

DISCUSSION

R. Thompson *(UK)*

Thank you. The discussion is now open on the papers of Drs. Snoad and Hedley.

D.A. Lawes *(UK)*

Could I ask either Dr. Snoad or Dr. Hedley if they could comment more specifically on pod set and pod shedding in this crop?

C.L. Hedley *(UK)*

Yes, there is shedding but probably not for the same reason as with *Vicia faba*. There tends to be a fall off thoughout the development of the plant in yield components. When we measured yield components such as the number of nodes producing pods and the number of pods per node etc., at the end of its development each yield component tended to be about half of its potential value. When these are multiplied up the result is 20% or less of what is admittedly an unrealistic potential. So there is a large reduction below the potential. Because flower and pod drop do occur we prefer the single podded or twin podded types, multi-podded peas being much more susceptible to pod and flower drop.

B. Snoad *(UK)*

I would like to agree with what Dr. Hedley has said although flower drop is not very much of a problem. Pod drop is - but only with an excessive number of flowers; in other words when intra-plant competition is excessive.

D.A. Lawes *(UK)*

Is there any difference in the level of pod dropping between the leafed pea and the leafless pea?

C.L. Hedley

Not as far as I can see.

G.G. Rowland *(Canada)*

What is the nature of the nitrogen fixation of the leafless pea?

C.L. Hedley

It is related to plant size. They fix nitrogen perfectly well. Individual plants fix less because they tend to be smaller than conventional types.

G.G. Rowland

Is fixation terminated at the same time as normal, about the start of flowering?

C.L. Hedley

I do not know if that is true; I tend not to believe it.

R.B. Austin *(UK)*

If you have to sow more seeds of the topless in order to get similar biological and grain yield as from normal types, there may be a penalty in terms of seed rate which could make it economically unsatisfactory.

C.L. Hedley

Since seed is bought by weight and sown by weight, and as we are suggesting the ideotype should have smaller seeds, then if the seed weight is halved we should have twice the number and thus the cost would be the same.

R.B. Austin *(UK)*

As you know, if you sow 100 seeds you cannot guarantee how many plants will actually grow. If with weak 'uncompetitive' plants, sown at 100 seeds/m^2 only 50 plants grow, then there would be a lower yield from the leafless types than from the leafy ones.

C.L. Hedley

That is not necessarily so because, irrespective of the competitiveness of the plant, the whole crop will ultimately stop growing anyway. If a less competitive plant was grown at a lower planting density, it may be that the plant would 'compensate' by growing for a longer period. Duration of growth is the key factor.

J. Picard *(France)*

I would like to underline that there is a difference between peas and many other legumes on the question of flower drop. In my opinion it has nothing to do with pollination. There are some peas where there is no flower drop, which contrast sharply with soyabeans, which are also autogamous and which show a high level of flower drop. An understanding of the physiological basis of this difference would be most valuable.

R. Thompson *(UK)*

I think there are differences between pea genotypes in the response of flower drop to environment. Physiological work on this has been done at the National Vegetable Research Station, Wellesbourne, in the UK and in Australia.

B. Snoad

As far as I am aware, the kind of work to which you are referring is with the vining pea, which is an entirely different kind of plant in its response to competition. Simultaneity of maturity is required in this crop and intra-plant competition is much greater than it would be in a dried pea crop.

C.L. Hedley

One of the other advantages of growing plants at high planting densities is that the variance ratio within the crop is increased as the planting density is increased, but there is a good chance that variation between crops will decrease and we may find more stable yields at high planting densities even though it will increase the variation within the crop.

M.C. Saxena *(Syria)*

Are there any estimates of photosynthesis of the pods for the leafless genotype?

C.L. Hedley

Yes. I do not think there is any difference in the way that pods behave in the leafed or leafless background. There is some suggestion that there is greater photosynthesis with the leafless but I would not expect this to be so.

R. Thompson

Before closing this discussion I would like to add one comment. The work obviously is novel and very stimulating. I realise there was insufficient time, but it would have been interesting to have heard how the responses of the leafless genotypes to population density and the interpretation of these responses correspond to the wealth of information that is already published on plant population; these genotypes are somewhat unique and it may be that their responses are quite different. We may have to look at them in a very different way to that of the normal genotype.

SOME REFLECTIONS AND DATA ABOUT USEFUL PLANT TYPES IN *Vicia faba* BREEDING

K. Nagl

Federal Institute for Plant Production and Seed Testing, Vienna, Austria.

ABSTRACT

Several mutant derivatives of cv. Kornberger Kleinkörnige (Vicia faba spp. minor) *selected for reduced apical dominance were analysed in respect of yield components and their interrelations. A marked reduction in height compared with the original indeterminate variety was associated with a high number of pod bearing stems and increased numbers of podbearing nodes and pods per plant. Numbers of seeds per pod were reduced but average seed weight remained unchanged. The high number of pod bearing nodes may provide a compensating mechanism to balance differences in population density. Compensation between the various yield components and the consequences of selection for high and low expression for any given component are presented in graphs.*

Changed plant architecture and responses in yield components may provide an additional insight into biotype performance.

CHANGES IN PLANT ARCHITECTURE AND RESPONSES IN YIELD COMPONENTS

Successful crop production depends on the effective exploitation of photosynthesis to achieve maximum biological yield. However in most cases, including field beans, only a part of the total crop - the seed - is of economic value. High economic yields require an effective distribution of assimilates at the right time.

In field beans indeterminant growth results in the apex remaining as a significant sink during and after the flowering period. This leads to competition between reproductive and vegetative organs, which may be overcome either by direct removal of the apex or by treatment with growth regulators, leaving the apex intact (Gehriger et al., 1979; Keller et al., 1980).

Breeding for reduced apical dominance is also of great interest as a means of changing plant architecture and providing a better supply of assimilate for pod set and grain development. Through breeding, an improved combination of characters is being sought either by promoting recombination and breaking linkages or utilising new gene sources by artificial induction. Such genotypes may contribute to a better understanding of biotype physiology and hence promote breeding progress (Chapman, 1979; Nagl, 1979).

Great interest has therefore been attached to studies on the topless mutant induced by Sjödin, which has a gene for terminal inflorescence (ti) and thus determinant growth (Sjödin, 1971). The lack of the vegetative top results in compensation by the leaves in the pod zone and probably no reduction in the supply of assimilate to the pods and seeds. A particular weakness of the original topless mutant and of its crosses is the low number of podbearing nodes per plant (Sjödin, 1977). The introgression of the terminal flowering allele in breeding for earliness affected maturity by about

10 days, and caused a reduction in yield of about 50%, accounted for mainly by reduced numbers of podbearing nodes. However, there is no evidence that genetic background significantly modifies the effect of the 'ti' allele on number of podbearing nodes (Aylmer et al., 1980).

On average there is also a lower number of seeds per pod in the topless crosses which may be related to a change in hormone balance due to the lack of an apex. There is a marked tendency to produce side shoots in later growth which may cause uneven ripening and reduce yield by upsetting the source/ sink relationship. Optimum growing conditions may overcome the reduced yields associated with these genotypes but the search for new gene sources for determinate growth is desirable. The present contribution will consider methods of evaluating varying degrees of reduced apical dominance.

Genetic control of the number of nodes/stem may indirectly determine both number of pods and number of leaves, the leaf/pod relationship being fixed. This model appears satisfactory for unbranched varieties, but in multibranched types the model would also need genetic control of branching. Early branching and more simultaneous flowering, together with an increase in podbearing nodes, might be considered as additional yield components. Components directly affecting seed yield are considered as first-order components e.g. pod number, seeds per pod and seed weight; second order components also influence yield along with numerous physiological processes.

The direct effect of a particular component upon yield can be largely compensated for by indirect effects of combinations of other components.

The principles relating to the balance between the various components should be given due consideration by the plant breeder. Selections should be evaluated under conditions that afford full expression of relevant genes, with continuing analysis of the initiation and interrelations of yield components.

Negative correlation among morphological components of yield is a widespread phenomenon and it raises the question of their biological origin and meaning. Negative correlations may arise in response to competition acting on developmentally flexible components. As the first component in the sequence of development uses greater or lesser amounts of assimilate, the next component in the sequence varies in growth accordingly. This can be interpreted as the consequence of sequentially developing components sharing a common pool of assimilate. Developmental plasticity of yield components could promote yield stability if variation in development of one component compensated for variation in another. Intense interplant competition for resources leads to intra-plant competition between reproductive structures and hence to compensation between components. Low to near zero correlations were found among components in non-competing spaced plants in contrast to that for closely spaced plants where negative correlations occurred. The consequences for the plant breeder of the fluctuating model, in addition to promoting recombination of unfavourably linked genes, would be to increase assimilate supply and reduce the capacity of a component to respond when resources are available. (Adams, 1967; Adams et al., 1971; Duarte el al., 1972).

Genes controlling the growth habit of determinate types may play such a role if they act like the dwarfing genes in wheat by lowering the endogenous gibberellin levels through inhibition (Baroncelli, 1980), so that the dwarf habit is retained even in conditions that usually promote stem elongation. Moreover, it may be supposed that a certain gibberellin with little influence on elongation may be efficient in flower initiation. This possible involvement of gibberellin in plant growth and development suggests that genes which control hormone synthesis may play a central role in developmental processes and might form the basis of useful studies on determinate types.

If strong linkages are prominent in establishing yield patterns the required flexibility of response would not be possible. If genes regulate the formation of one component without directly regulating one of the other components, a genetically independent influence on yield is indicated. This would result in component correlations generally near to zero in non-competitive or non-stress situations.

REDUCED APICAL DOMINANCE AND BRANCHING

The screening for genotypes with different degrees of reduced apical dominance by continuous reselection in a mutant population of cv. Kornberger Kleinkörnige resulted in a number of lines differing in plant height. These lines were compared with the indeterminate original variety Kornberger Kleinkörnige on a single plant basis for different yield components. Space planting and irrigation during the flowering period contributed to the input conditions that allowed full expression of relevant genes for each yield component.

On average, the selected lines were reduced in plant height by 40%. Since apical dominance was reduced, the tendency to produce side shoots was stimulated and values reached three times those of the indeterminate variety (Table 1). The most obviously changed component of this breeding material was the number of podbearing stems and might be of interest through its effects on reproductive development. A correlation matrix for 700 inbred lines (Frauen, 1977) showed negative correlations between number of flowering tillers and the majority of yield components analysed. In this work nine of thirteen correlations were negative and only two positive correlations were found, these being between number of tillers and seed weight and between number of tillers and yield.

Another yield component with the same high number of nine negative correlations with other components was seed weight. In the sequential pattern of development of yield components, number of stems occurs at the beginning and average seed weight is at the end. In highly fertile growing conditions, Thompson (1979) found that storage and structural material of the stem accounted for a considerable proportion of the additional

TABLE 1

MEAN VALUES OF FIRST AND SECOND ORDER YIELD COMPONENTS OF THE INDETERMINATE CV. KORNBERGER KLEINKÖRNIGE AND OF 23 MUTANT DERIVATIVES SELECTED FOR REDUCED APICAL DOMINANCE.

	No. of stems	Plant height cm	No. of podbear. nodes	No. of pods	Grain yield per plant g	No. of seeds per pod	No. of pods per node	Average seeds weight g	Harvest index
cv. Kornberg Kleinkörnige indeterminate	1.3 ±0.10	110.0 ±5.7	14.7 ±0.67	28.9 ±1.85	37.3 ±2.7	2.75 ±0.17	1.96 ±0.07	440.3 ±14	0.49 ±0.015
23 lines with reduced apical dominance	4.2 ±0.16	65.6 ±3.5	22.7 ±1.35	36.4 ±2.2	31.1 ±2.4	1.95 ±0.08	1.62 0.03	450.1 ±18.6	0.38 0.01
Range	3.0-6.3	44-97	7-38	19-52	11-59	0.7-2.7	1.4-1.9	274-611	0.26-0.49
Coefficient of variance for lines	0.19	0.27	0.29	0.30	0.39	0.21	0.08	0.20	0.17

assimilate produced. This may partly explain the competition for assimilates with other components that is reflected in the high number of negative correlations. On the other hand, stems may function as a store, redistributing the accumulated assimilates later.

The marked reduction in height of the mutant derivatives was associated with high numbers of podbearing stems with an average of 4.2 and a range of 3.0 - 6.3 stems for single lines. Significant negative correlations were limited to two components of thirteen components analysed, and these were grain weight (r = -0.42*), and grains/pod (r = -0.41*). Apart from the significant positive correlation between plant height and number of podbearing nodes (r = 0.51*) the number of stems also contributed positively to both number of podbearing nodes (r = 0.51*) and the number of pods (r = 0.55*). This is confirmed by the positive partial correlations between these two components and the number of stems, eliminating the influence of plant height (Figure 1).

The contribution of branching to seed yield might be evaluated differently. Both unbranched and multi-branched types are integrated into commercial varieties of different leguminous crops such as soyabeans, peas and Phaseolus beans.

In a few crops canopy structure can be controlled by planting density but usually branching and leaf production eventually compensate for changes in density. Therefore, suggestions for obtaining higher and more stable yields of field beans include the development of a plant type with few stems which branch simultaneously from the basal nodes at an early stage. This would lead to greater synchrony of development and avoid the effects of failure of later formed sinks. Where several sinks are competing for a limited supply of assimilate, sink size may be a significant control mechanism, partitioning tending to favour the larger sink. The relative strength of the sinks largely determines the pattern of assimilate

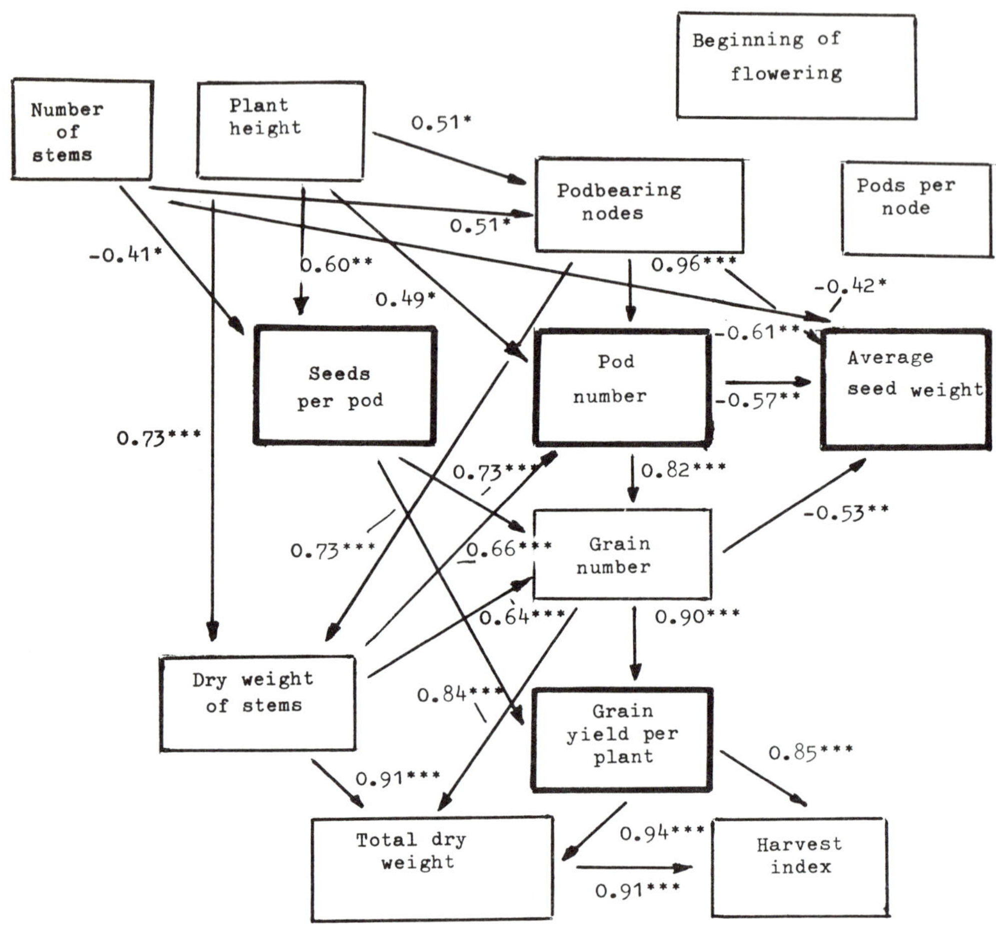

Fig. 1 Some interrelations of yield components in *Vicia faba* lines selected for reduced apical dominance.

movement and thus there is no especially preferred position for storage organs.

There may well be situations where neither source nor sink limits the rate of accumulation of assimilate and where the capacity of the translocation system may be limiting at some point between the source and sink. Gibberellic acid (GA) increases either leaf or stem growth of bean but not usually both, which could imply competition for growth substances or photosynthates. By removing the apical growing point in Dwarf French Bean plants, Wheeler and Humphries (1964) tried to eliminate this source of competition but found that inability of GA to move freely was more important than competition for its supply.

The pattern of assimilate distribution changes as plants grow and develop leaves and new sinks. Dwarf pea varieties tend to throw unequal shoots, the less vigorous of which senesce early whereas tall varieties generate shoots more equal in size and which often survive to maturity (Pate, 1975).

In soyabeans, nodes which develop simultaneously on the main stem and branches (co-growing nodes) tend to develop similarly. They bloom at the same time and have similar flower numbers, pod numbers and abortion percentages. Soyabeans may provide instructive plant models for the determinate plant type of field bean and their patterns of branching. Two types of growth habit are common; there are determinate types in which the stem and branches possess a terminal raceme with a cluster of pods, and there are the indeterminate types. Additionally there are two genetic types with the determinant habit, i.e. dt_1 recessive and dt_2 dominant to indeterminancy. Determinancy reduces plant height and number of nodes more markedly in dt_1 plants and these also branch more than dt_2 types. Stem growth of dt_1 terminates either before flowering or shortly thereafter, whereas dt_2 is intermediate to indeterminate in habit. Sequence of flowering differs with growth habit. In dt_1 determinates, length of flowering is associated with duration of flowering at a single node whereas, in indeterminate and dt_2 determinates, it depends upon sequental flowering of the nodes. In indeterminate cultivars there is much greater seed production from centre nodes whereas, in dt_1 determinates, podding is heavier in the upper half of the plant, and branches and the terminal racemes make a significant contribution (Shibles et al., 1975). Some similarities exist between our different lines of field bean with reduced apical dominance or of determinate habit and the terminal types of soya. In general the majority of the lines branch at an early stage which leads to uniform stem development and uniform distribution of pods between stems. Determinate types of field bean with terminal flowers and a more positive termination of stem growth are similar to dt_1 types of soya in relation to sequence of flowering and pod set. In addition, larger leaves characterise this type of field bean, which is a feature also found in determinate soya. Intermediate types of field bean without terminal flowers concentrate podset on the lower parts of the stems, which corresponds to the dt_2 determinates of soya.

CONSEQUENCES OF REDUCED APICAL DOMINANCE ON REPRODUCTIVE DEVELOPMENT

In contrast to the findings of Sjödin (1977) for the topless mutant and its derivatives, the material described here had 50% more podbearing nodes than the orginal indeterminate variety and 26% more pods per plant (Table 1). The number of podbearing nodes may provide a major compensating mechanism tending to balance differences in population density. Stabilisation of a higher number of podbearing nodes may also lead to improved harvest index indicated by their positive correlation (r = 0.49*).

Besides a higher number of pods (+ 26%) which was associated with an increased number of podbearing nodes (r = 0.96*** Figure 1) there was a reduction in seeds per pod of about 30% (Table 1). Seeds/pod was negatively correlated with numbers of stems / plant (r = -0.41*) but was not associated with changes in average seed weight. A similar response resulted from decapitation where reduced pod drop increased the number of pods, but the supply of assimilate for grain filling was insufficient to meet demand, so that both the number of grains per pod and grain yield were lower than for the control. The higher number of pods which resulted from increased podding nodes, in lines selected for reduced apical dominance, led to strong competition and reduced numbers of seeds/pod, and also reduced total grain yield. However, a high level of variability in numbers of beans/pod suggest good prospects for efficient selection for yield increase as shown in Figure 3. Average seed weight is another important yield component and on average was not much affected, but of all the components examined it showed the greatest number of negative correlations to other components; this agrees well with the correlations shown by Frauen (1977). In the present breeding material, only six negative correlations were calculated between different components, four of them were connected with average grain weight. This is a very low proportion of the total number of 78 correlations and may reflect the non-competitive growing conditions under which the lines were tested so as to obtain maximum expression of relevant genes.

Selection for high and low levels of expression of the various yield components may give some idea of the interaction between the selected yield component and the rest. Modification of a basic pattern can result from difference in variability, to genetic correlations, or from developmentally induced relationships and from compensating mechanisms.

Figures 2 - 4 show the relationships between a number of yield components for the two selections with the highest and lowest values for each yield component in turn. For example, the first graph shows relative values of nine components of yield for the two lines with the highest and lowest values for numbers of stems per plant. The second graph shows the values for the same nine components but for two lines which may or

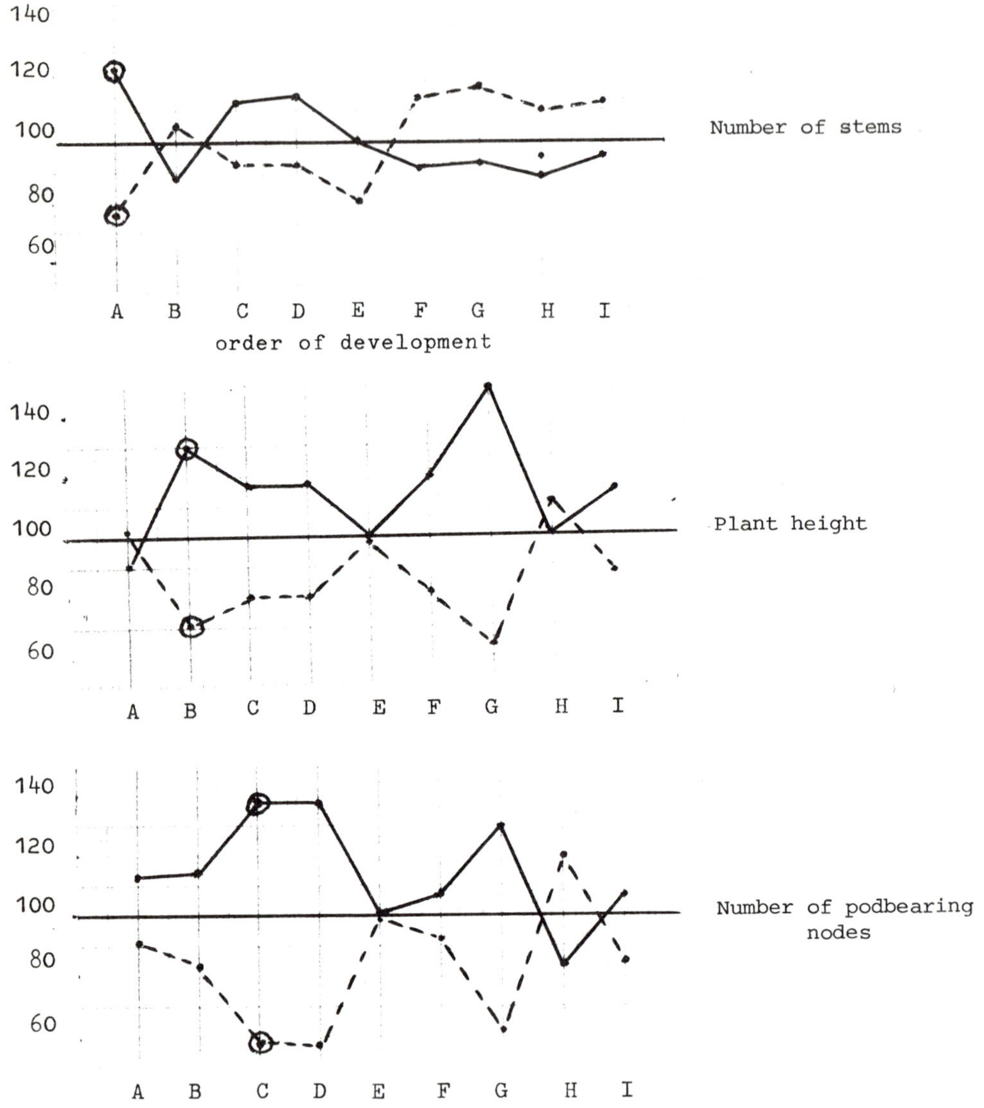

Yield components:

A Number of stems
B Plant height
C Number of podbearing nodes
D Number of pods
E Pods per node
F Seeds per pod
G Grain yield per plant
H Average grain weight
I Harvest index

Fig. 2 Compensatory fluctuations of different yield components from the mean of 23 *Vicia faba* lines reduced in apical dominance shown by high and low selections.

% of mean

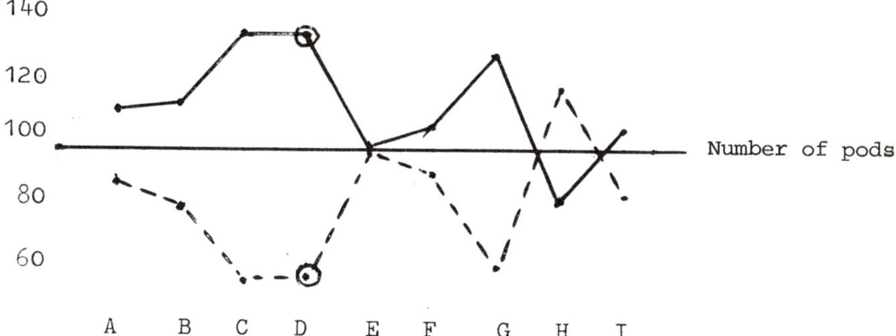

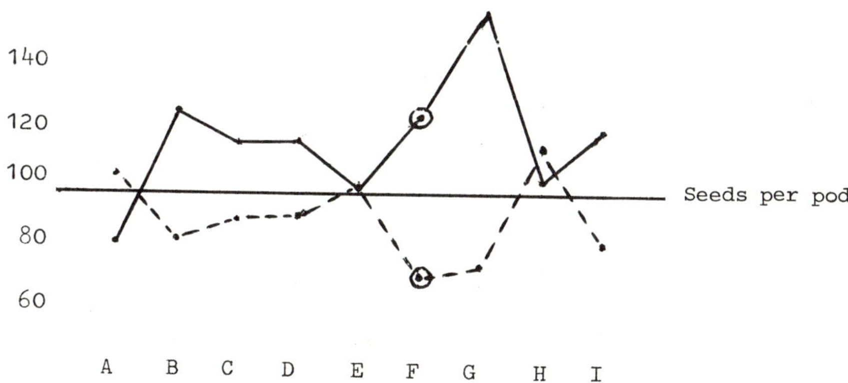

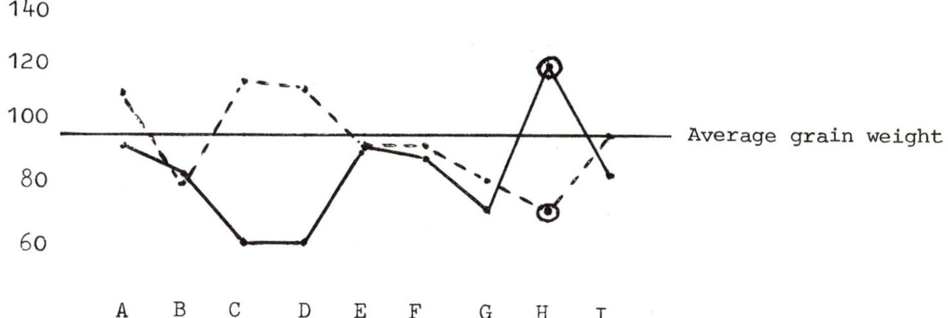

Fig. 3 Compensatory fluctuations of first-order yield components. For legend see Figure 2.

% of mean

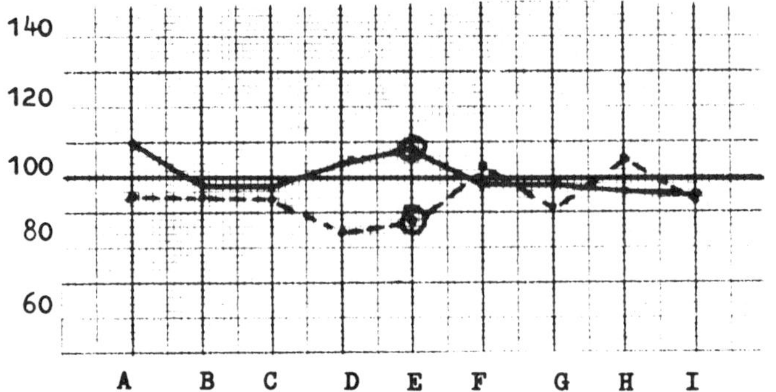

Pods per node

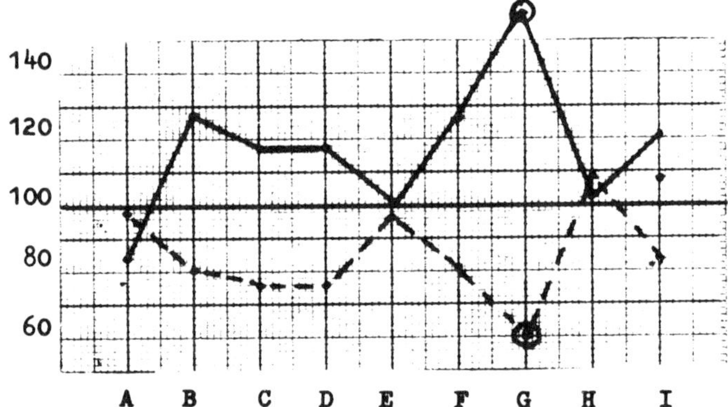

Grain yield per plant

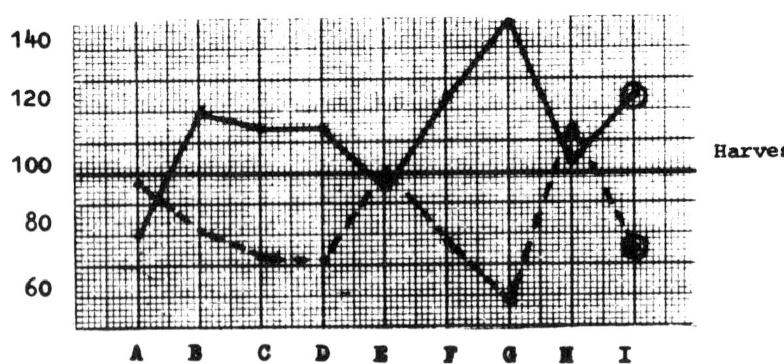

Harvest index

Fig. 4 For legend, see Figure 2.

may not be the same as for other graphs, but which had the maximum and minimum values for plant height. And so on through the rest of the nine components. Presented in this way, it is easy to see the consequences on other components of selecting lines, within the selections for reduced apical dominance, with extreme values for any given component of yield.

Compensation by one component for another can be substantial and different genotypes may respond differently in relation to available resources.

The additional genetic and phenotypic diversity resulting from mutation and recombination has led to a better understanding of biotype performance.

REFERENCES

Adams, M.W., 1976. Basis of yield component compensation in crop plants with special reference to the field bean *Phaseolus vulgaris*. Crop Science 7, 505-510.

Adams, M.W. and Grafius, J.E., 1971. Yield component compensation - alternative interpretations. Crop Science 11, 33-35.

Aylmer, J.M. and Walsh, E.J., 1980. An evaluation of two approaches to breeding for earliness in field beans (*Vicia faba* L.). Irish Journal of Agr. Res. (in press).

Baroncelli, S., Buiatti, M. and Magnani, G., 1980. Control of gibberellin action by 'Semidwarf' genes in Durum wheat. Zeitschrift für Pflanzenzüchtung 84, 219-225.

Chapman, G.P., 1979. New concepts in breeding high yield *Vicia faba* L. Some current research of *Vicia faba* in Western Europe. Commission of the European Communities, EUR 6244 EN. 293-302.

Duarte, R.A. and Adams, M.W., 1972. A path coefficient analysis of some yield component interrelations in field beans (*Phaseolus vulgaris* L.). Crop Science 12, 579-582.

Frauen, M., 1977. Yield and yield components of different lines of *Vicia faba*. Meeting of the EC-working group on seed legumes Stuttgart-Hohenheim and Zürich, 42-48.

Gehriger, N. and Keller, E.R., 1979. Einfluss des Köpfens der Triebe auf die Versorgung der Blüten bei der Ackerbohne (*Vicia faba* L.) mit 14 C. Schweiz. Landwirtschaftl. Forschung 18, 1/2, 59-80.

Jaquiéry, R. and Keller, E.R., 1978. Beeinflussung des Fruchtansatzes bei der Ackerbohne (*Vicia faba* L.) durch die Verteilung der Assimilate Teil I. Angew. Botanik 52, 261-267.

Keller, E.R. and Bellucci, S., 1980. Effect of growth regulators on faba bean (*Vicia faba* L.) development. Fabis Newsletter, 2, 36.

Nagl, K., 1979. Results of mutation and breeding work in *Vicia faba* in Austria. Commission of the European Communities EUR 6244 EN, 355-369.

Pate, J.S., 1975. Pea. Crop Physiology, Cambr. Univ. Press, 191-224.

Shibles, R.M., Anderson, I.C. and Gibson, A.H., 1975. Soyabean. Crop Physiology. Cambr. Univ. Press, 151-190.

Sjödin, J., 1971. Induced morphological variation in *Vicia faba* L. Hereditas, 67, 155-180.

Sjödin, J., 1977. The determinant's growth habit - a promising goal in breeding *Vicia faba*? Meeting of the EC-working group on seed legumes Stuttgart-Hohenheim and Zürich, 67-75.

Thompson, R., 1979. Crop growth and partition of assimilates in field bean *(Vicia faba)*: Responses to elimination of some major constraints. Commission of the European Communities EUR 6244 EN, 407-420.

Wheeler, A.W. and Humphries, E.C., 1964. Separation of effects of gibberellic acid on leaf and stem growth of Dwarf French Bean. Nature Vol 202, 616.

ASPECTS OF FABA BEAN IDEOTYPES FOR DRIER CONDITIONS

M.C. Saxena, G.C. Hawtin and H. El-Ibrahim
ICARDA, Aleppo, Syria.

ABSTRACT

The faba beans in West Asia and North Africa are exposed to cold winters with uncertain rainfall, short springs and a hot and dry summer. Studies at the ICARDA site in North Syria suggest that the faba bean plant ideotypes suitable for these conditions should have the following attributes: ability to grow at low temperatures and recover from frost damage during the early vegetative growth; rapid reproductive growth in spring to ensure yield build-up before the hot summer sets in; moderate branching with 3 - 5 main stalks/plant and a leaf area index (LAI) of about 5.5 in the normal stands of 20 - 30 plants/m^2*; a deep root system capable of extracting moisture from deeper layers; high auto-fertility; and resistance to common diseases and pests, particularly* Orobanche.

INTRODUCTION

The region of West Asia and North Africa is characterised by a typical 'Mediterranean-type' climate - winter rainfall alternating with summer drought. Considerable climatic variability occurs throughout the region because of the influence of the major geographical features. The bulk of the cropping area in the region is located on the low-lands where both moisture supply and temperature exercise great influence on crop productivity.

The rainfall, which is highly variable in terms of both magnitude and distribution, is restricted to the period from October to June. Most of the cropped area receives an annual rainfall varying from 200 - 500 mm between years. Low temperatures early in the season retard early crop growth and the crop is exposed to a frost risk throughout this period with the late frost risk extending up to the end of March or the beginning of April (Harris, 1979). With the onset of spring, the maximum temperature increases very rapidly, attaining values well above 30°C, and hastens crop maturity. The grand period of growth occurs during a period of drought and so with increased evaporative demand of the atmosphere and dwindling soil moisture supply. An example of this type of environment is shown in Figure 1, which represents weather data for Tel Hadia, North Syria for 1977/78.

Faba beans which are an important source of edible proteins in the region, are planted in winter generally in low-land areas where either annual precipitation is high (above 600 mm) or irrigation facilities exist. With consumer demand and market prices for the faba bean rising, the need for extending its cultivation to drier areas has been increasing. Even in the high rainfall areas, the stability of yield is reduced because of occasional periods of drought. Development of genotypes with superior performance under drier conditions, therefore,

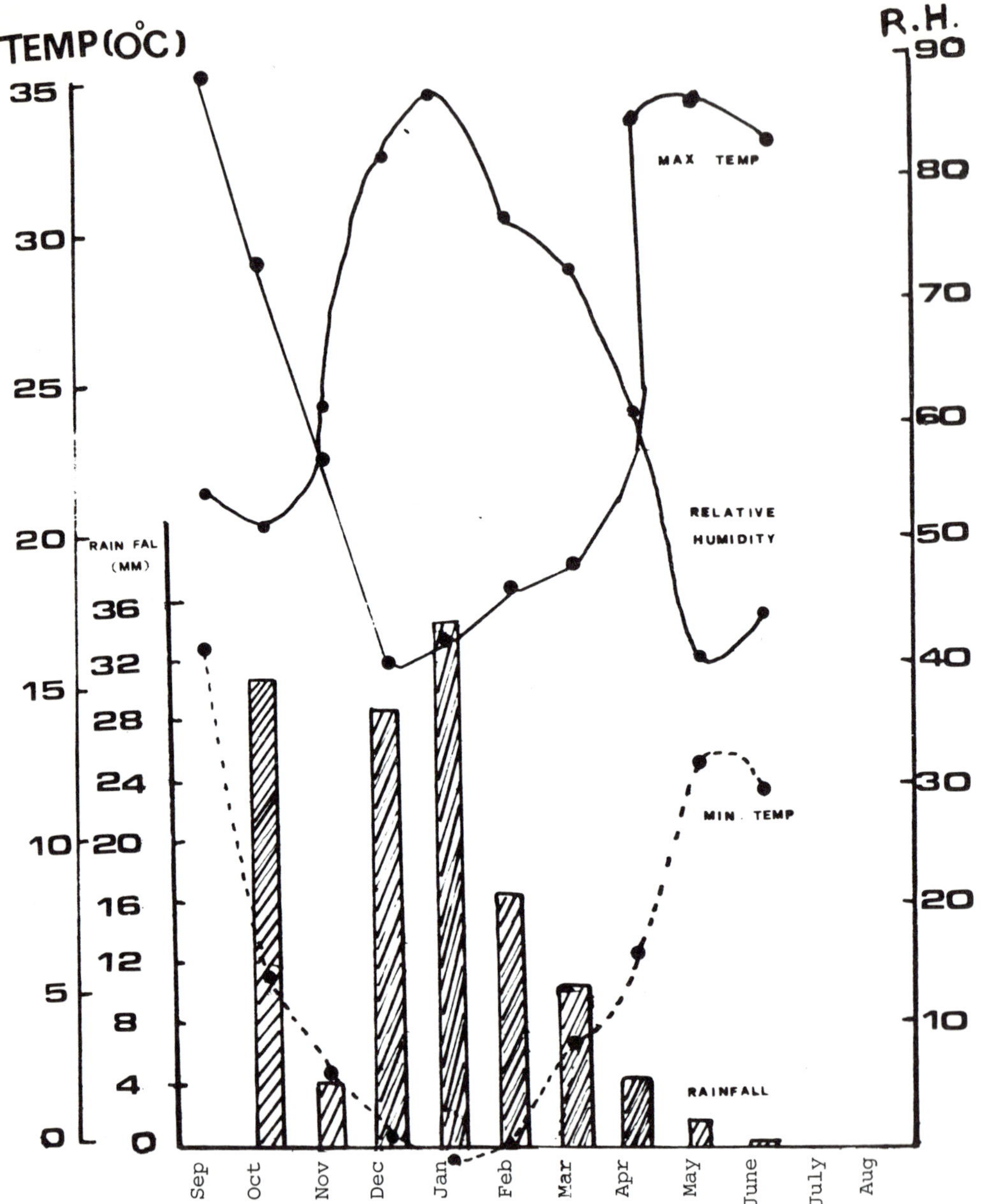

Fig 1. Weather conditions during the faba bean growing season at Tel Hadia, North Syria, 1977-78.

is an attractive proposition. Some aspects of appropriate plant ideotypes for such conditions are discussed below.

DROUGHT AVOIDANCE OR TOLERANCE

For making best use of limited rainfall in the drier areas, it is necessary that the crop be planted early in winter, when most of the precipitation occurs. Date of planting studies have shown (Table 1) that drastic reductions in the yield of most of the genotypes occur as the planting is delayed beyond November. This is particularly true for European cultivars, which may fail to make satisfactory reproductive growth if planting is delayed beyond December. This requirement for planting in early winter necessitates that there should be a fair measure of tolerance to low temperature in the genotypes, not only to maintain a satisfactory rate of growth but also to survive the occasional frost that may occur during this period. Figure 2 shows the weather conditions in Tel Hadia from November to January during 1978/79 and 1979/80. The latter cropping season had repeated cycles of frost at the end of November and December and the beginning of January, which caused considerable damage to the faba beans. Damage varied not only with the date of planting but also with different cultivars. The earlier the planting, the greater was the damage, (Table 2). Locally well adapted genotypes such as Syrian Local Large, Syrian Local Small, Aquadulce and New Mammoth showed much less damage than the other less well adapted genotypes (i.e. Express, Giza-2, Giza-3 and Hudeiba-72). One of the local genotypes, a landrace from the coastal area of north Syria, also showed a high susceptibility to cold. The well adapted genotypes showed a high degree of recovery at the end of the frost period, whereas the susceptible ones, such as Hudeiba-72 from Sudan, suffered severe injury and failed to recover.

Drought avoidance and/or tolerance is another important character required to give stable yields in dry areas of agriculture. The chances of avoiding drought are greater if the crop is able to attain a satisfactory rate of growth during

TABLE 1

EFFECT OF DATE OF PLANTING ON THE YIELD (kg/ha) OF SOME FABA BEAN GENOTYPES AT TEL HADIA, N. SYRIA.

Date of planting	Syrian local Large		Aquadulce		Maris Beagle	Three fold white	
	1977-78	1978-79	1977-78	1978-79	1977-78	1977-78	1978-79
Oct. 30	-	3159	-	2852	-	-	938
Nov. 29	-	2640	-	3514	-	-	771
Dec. 4	1481	-	1148	-	151	149	-
Dec. 24	-	1757	-	1996	-	-	696
Jan. 14	667	-	773	-	0	40	-
Jan. 25	-	650	-	921	-	-	161
Feb. 2	327	-	398	-	0	55	-
March. 10	0	-	0	-	0	0	-

TABLE 2

FROST DAMAGE IN SOME SELECTED VARIETIES OF FABA BEAN AS AFFECTED BY DATE OF PLANTING IN 1979-80 AT TEL HADIA, N. SYRIA.

Genotypes	% Plants from different planting dates				
	23.9.79		11.11.79		4.12.79
	damaged by Dec.31	killed by Mar.18	damaged by Dec.31	killed by Mar.18	killed by Mar.18
Syrian L. Large	70	13.2 ± 3	3	3.6 ± 4	0
Syrian L. Medium	72	9.8 ± 4	0	0.6 ± 4	0
Aquadulce	75	3.8 ± 4	0	3.6 ± 4	0
Lattakia L. Large	85	35.1 ± 11	3	1.9 ± 2	0
New Mammoth	65	5.3 ± 2	8	1.4 ± 2	0
Express	95	70.6 ± 9	10	4.5 ± 3	4.9 ± 5
Giza-2	97	71.3 ± 9	18	17.6 ± 4	0
Giza-3	100	75.4 ± 5	68	33.7 ± 13	2.4 ± 4
Hudeiba-72	100	100.0 ± 0	90	85.7 ± 8	27.5 ± 9

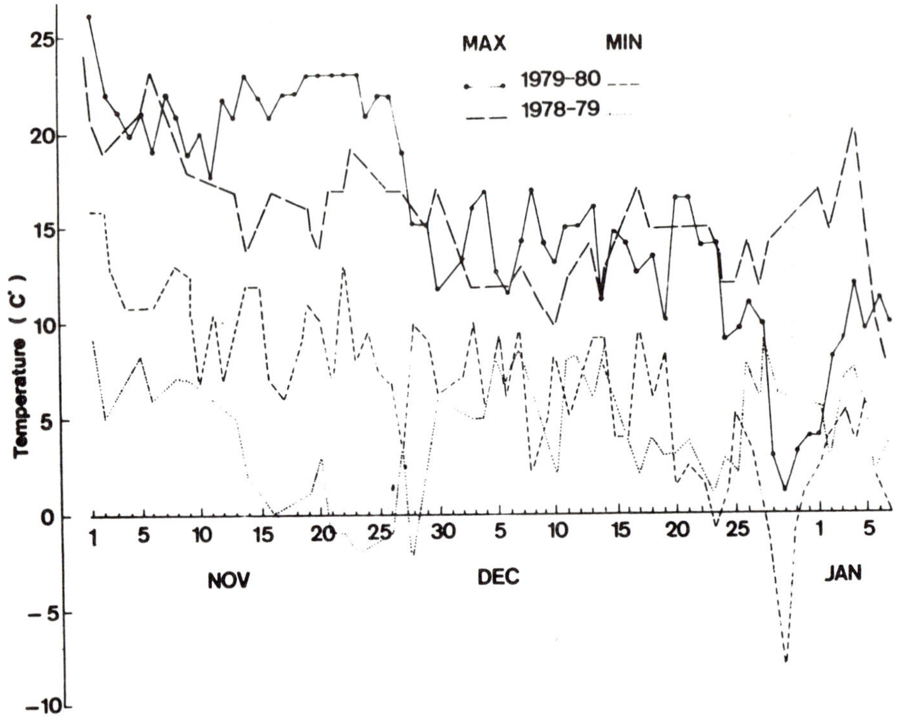

Fig 2. Change in the maximum and minimum temperature during the early growing season of faba beans at Tel Hardia, North Syria, in 1978-79 and 1979-80.

the winter, enabling the plant to develop a vegetative frame capable of making best use of the environment during the short spring. Early maturity would, therefore, be a desirable attribute to look for. In addition a certain degree of drought tolerance would also be necessary. In view of the lack of correlation between various tests for drought tolerance and actual field performance under low moisture supply condition, screening of faba bean genotypes at ICARDA has been done by the latter method using sites with different levels of rainfall. During 1978/79 the seasonal precipitation at Tel Hadia was a little less than 250 mm. In this season several genotypes (including the genotypes 1, 2, 3, 4, 8 and 9 in Table 3) yielded more than 2 000 kg/ha. Their evaluation at several rainfed sites

TABLE 3

PERFORMANCE OF SOME SELECTED GENOTYPES OF FABA BEAN AT DIFFERENT LOCATIONS IN NORTH SYRIA DURING 1980 CROPPING SEASON

S.No.	Genotypes	Tel Hadia					Jinderis		Brida	
		Irrigated		Rainfed (427.7 mm)*			Rainfed (376.7)*		Rainfed (268.7)*	
		Yield (kg/ha)	Rank	Yield (kg/ha)	Rank	Water use from 19 Feb to 21 May (cm)**	Yield (kg/ha)	Rank	Yield (kg/ha)	Rank
1	78-S-49907	2784	2	2689	2	25.72	2281	4	978	1
2	78-S-48821	2200	8	2577	3	24.93	2247	5	650	10
3	78-S-49395	2436	4	2265	7	24.66	1933	7	745	8
4	78-S-49694	2707	3	2277	8	24.09	2533	3	811	5
5	Seville Giant	2419	5	2298	6	23.94	1457	10	774	6
6	Syr.L. Large	2964	1	2713	1	25.53	2790	2	826	3
7	Leb.L. Small	1737	10	2458	5	24.53	1762	9	760	7
8	78-S-49892	2377	7	2114	9	25.33	1871	8	877	2
9	78-S-49896	2418	6	2482	4	24.72	3014	1	814	4
10	Giza-3	1733	9	1881	10	23.98	2190	6	721	9
	L.S.D. 5%	Genotypes = 285; GXM.S. = 403					N.S.		274	
	C.V.	10.24%					29.3%		20.10%	

* Total seasonal rainfall (mm)

** Total seasonal water use for S.L. large faba bean in separate trial = 384.5 mm.

during the 1979/80 season revealed that they tended to maintain their superiority over other genotypes which were not so well adapted to the drier conditions. Preliminary studies on the soil moisture extraction from different profile depths by the use of a neutron probe indicated that one of the above selections (78-S-49907) extracted moisture from deeper layers than Seville Giant and Giza-3.

YIELD COMPONENTS IN RELATION TO DROUGHT

In an attempt to identify the yield attributes that may be responsible for superior performance under rainfed conditions, the yield and yield attributes of 3 top-ranking and 2 bottom-ranking entries in a large seeded international yield trial (BIYT-80) were compared under rainfed and irrigated conditions (Table 4). The two sets of trials were located on different but nearby fields at Tel Hadia. The ratio of rainfed yield to irrigated yield was higher in the entries ranking higher in the yield performance under rainfed conditions. No single attribute of yield could be identified to which this superiority in yield performance could be attributed. A higher number of branches/plant, which in turn resulted in increased pods/plant, was important. But the ability to retain a high seed number/pod and 100 seed weight under rainfed conditions was also of significance.

The course of leaf area build-up in the local, well adapted, large seeded cultivar is shown in Figure 3. From the onset of flowering until pod development, there was a rapid increase in leaf area index. The significance of variation in the leaf area index (LAI) at different stages of growth was evaluated in a trial, the results of which are presented in Table 5. It appears that, whereas defoliation during the vegetative stage was of little significance, defoliation at the pod developing stage tended to cause yield reduction. Opening up the canopy by partial defoliation in the upper half of the plant at the flowering stage, which resulted in a LAI of 5.34 at the pod development stage, gave the highest yield, although this was

TABLE 4

PERFORMANCE OF SOME SELECTED ENTRIES IN BIYT-L-80 UNDER RAINFED AND IRRIGATED CONDITIONS AT TEL HADIA, N. SYRIA

Genotype	ILB no.		Grain Yield (kg/ha)		Branches/ plant		Pods/ plant		Seeds/ pod		100 Seed weight (g)	
			R	I	R	I	R	I	R	I	R	I
39MB (Syria)	1799	A	2612(1)	4208(4)	3.05	3.45	40.2	49.5	2.39	2.38	125.6	147.6
		B	0.62		0.88		0.81		1.00		0.85	
Eleg.5MC$_1$	1805	A	2607(2)	4185(5)	2.45	3.55	29.2	51.0	2.62	2.78	126.3	118.5
		B	0.62		0.69		0.57		0.94		1.06	
Local Check 'A'		A	2590(3)	3729(13)	3.00	4.55	31.2	54.7	1.74	2.17	166.7	151.8
		B	0.69		0.67		0.57		0.80		1.10	
Reina Blanca	1270	A	2242(12)	4075(6)	2.60	3.40	26.2	37.2	3.66	3.87	100.4	121.0
		B	0.55		0.76		0.70		0.94		0.82	
Lattakia local	1815	A	1839(16)	4445(1)	2.50	3.40	31.5	37.7	2.53	2.96	119.0	139.3
		B	0.41		0.73		0.83		0.85		0.85	
L.S.D. 5%			338	710								
C.V.			10%	12.5%								

R = rainfed; I = irrigated; A = absolute; B = R ÷ I.
Number in parenthesis shows rank

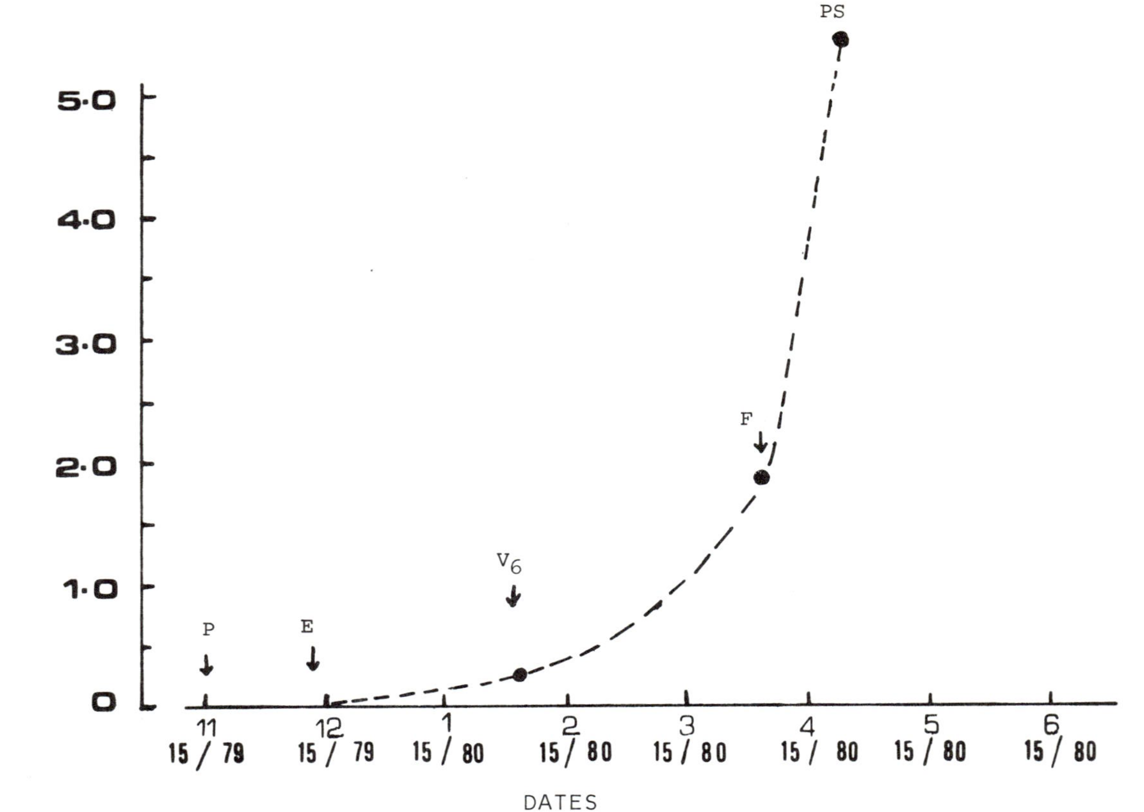

Fig 3. Leaf area index (LAI) of Syrian Local Large faba bean at various stages of growth in a normal stand of 20 plants per m^2. P = planting, E = emergence, V_6 = stage with six vegetative nodes, F = 100% flowering stage, PS = pod development stage.

TABLE 5

YIELD OF SYRIAN LOCAL LARGE (ILB-1814) FABA BEANS AS AFFECTED BY DEFOLIATION AT VARIOUS STAGES OF GROWTH

Treatment	LAI* at pod set	Grain yield (kg/ha)
1. Control	6.48	3744
2. 33% defoliation at V.6 stage	4.17	3578
3. 50% defoliation at V.6 stage	4.69	3429
4. 33% defoliation at 100% flowering	6.03	3228
5. 50% defoliation at 100% flowering	4.03	3157
6. 50% defoliation in upper half of the canopy at 100% flowering	5.34	3859
7. 33% defoliation at the appearance of at least one fully expanded pod (pod set)	3.08	3045
8. 50% defoliation at pod set	2.41	3217
9. 50% defoliation in the upper half of the canopy at pod set	4.33	2828
10. Decapitation of the plants above 5 flowering nodes at pod set	2.86	2316
L.S.D. 5%		677
C.V.		14.40%

*Leaf area was measured by using whole leaf (leaflets + rachis), after completion of the defoliation treatment.

not significantly different from the yield of the control. A LAI of 5.5 seems to be optimum for this crop, which was raised with one supplemental irrigation in the post rainy period (Figure 4).

The significance of different degrees of branching and determinate growth habit for a dry environment has been studied at the ICARDA site in Tel Hadia by applying debranching and detopping treatments to the 'Syrian Local Large' landrace. Four plant types were achieved: normal moderately branching plant with indeterminate growth (T_1); one main stalk only retained per plant by repeated debranching (T_2); three primary stalks retained per plant by debranching (T_3); determinate growth habit simulated by detopping all the shoots at the third node from the top after anthesis occurred on the first three flowering nodes (T_4). The performance of these plant types was studied at five levels of plant population, viz. 8, 16, 24, 32 and 40 plants/m^2 obtained by between-plant spacings of 25, 12.5, 8.3, 6.2 and 5.0 cm in rows 50 cm apart. The results of these studies are presented in Tables 6, 7, 8 and 9.

The number of branches/plant (Table 6) in the control (T_1) ranged from 5.7 to 3.2 depending on population density, whereas in the detopping treatment it ranged from 8.4 to 4.3. Apparently the detopping treatment encouraged the development of additional branches late in the season. The LAI at the 3-flowering node stage (Table 6) was highest in the control and smallest in T_2, increasing with increased population density in all treatments. The difference in LAI was small between T_1 and T_3.

Grain yield (Table 7) was highest in the control (T_1), which was similar to that that of T_3, but was significantly lower in T_2 and T_4. Even at increased population density the yield in T_2 (mono-stalked plant) and T_4 remained lower than the yield in T_1 and T_3.

The effect of detopping and debranching on the number of pods/plant (Table 7) showed the same trend as that for grain

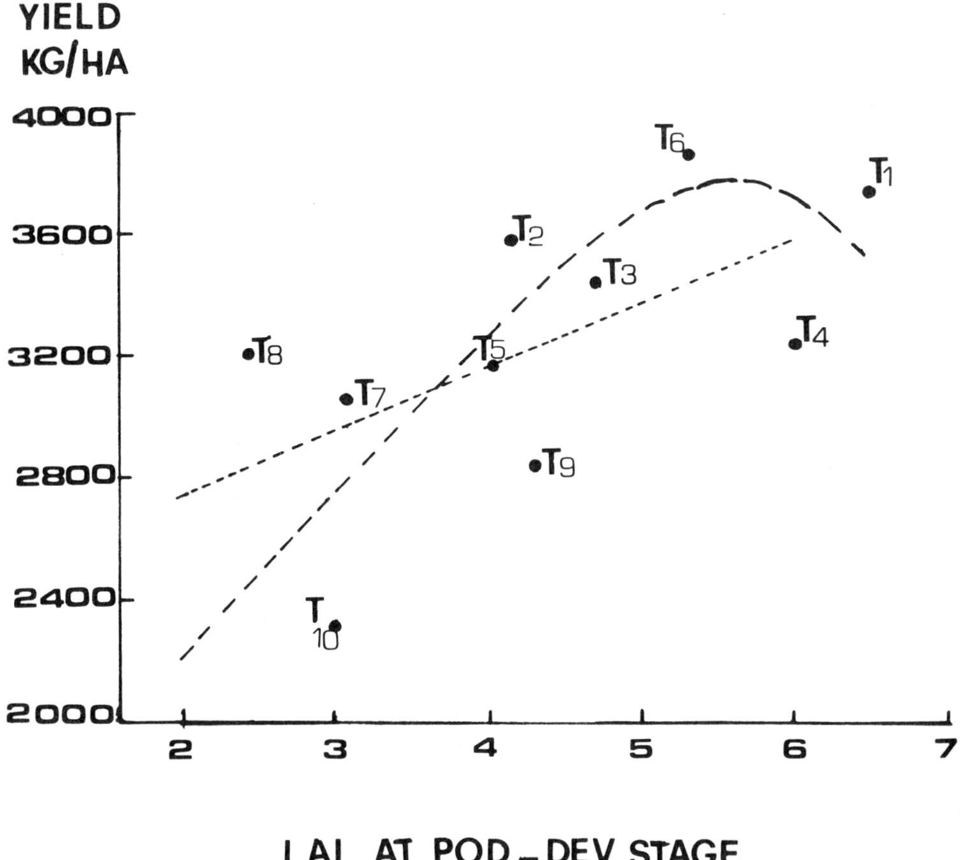

Fig 4. Relationship between leaf area index (LAI) at pod development stage and the seed yield of Syrian Local Large faba bean at Tel Hadia, North Syria, 1979-80.

TABLE 6

EFFECT OF DEBRANCHING AND DETOPPING TREATMENTS AT THE THREE FLOWERING NODE STAGE ON THE LAI AND NUMBER OF PRIMARY STALKS PER PLANT

Population per m^2	LAI				Primary stalks/plant at maturity			
	T_1	T_2	T_3	T_4	T_1	T_2	T_3	T_4
8	2.59	1.51	3.31	2.55	5.7	1	3	8.4
16	3.44	3.11	4.00	3.62	4.2	1	3	8.4
24	5.46	4.08	5.44	4.58	4.4	1	3	5.8
32	7.57	4.51	6.40	4.93	3.2	1	3	4.9
40	8.86	4.98	8.41	6.63	3.5	1	3	4.3

TABLE 7

EFFECT OF DEBRANCHING AND DETOPPING OF SYRIAN LOCAL LARGE FABA BEAN ON THE GRAIN YIELD AND PODS/PLANT AT VARIOUS LEVELS OF PLANT POPULATION

Population per m^2	Grain yield (kg/ha)					Pods/plant				
	T_1	T_2	T_3	T_4	Mean	T_1	T_2	T_3	T_4	Mean
8	2605	1823	2492	2126	2261	17.3	9.1	15.8	13.0	13.8
16	3762	2588	3288	3072	3178	11.9	10.0	11.4	9.8	10.8
24	3749	3411	3668	2956	3446	8.6	7.5	11.7	6.5	8.6
32	3801	3301	3541	3124	3441	9.0	6.7	8.6	6.5	7.7
40	3549	3374	3439	2917	3320	8.9	5.4	6.5	4.9	6.5
Mean	3493	2899	3286	2839		11.2	7.7	10.8	8.1	
	Population	Treat.	PXT			Population	Treat.	PXT		
L.S.D. 5%	331.3	242.1	N.S.			1.28	1.25	2.8		
C.V.	13.7%	12.1%				17.5%	20.8%			

TABLE 8

EFFECT OF DEBRANCHING AND DETOPPING TREATMENTS ON TOTAL PODS AND POD BEARING NODES/PLANT OF SYRIAN LOCAL LARGE FABA BEAN

Population per m^2	Total pods/plant				Total pod bearing nodes/plant			
	T_1	T_2	T_3	T_4	T_1	T_2	T_3	T_4
8	17.3	9.1	15.8	13.0	16.4	7.5	14.3	11.8
16	11.9	10.0	11.4	9.8	11.7	8.3	10.9	9.7
24	8.6	7.5	11.7	6.5	8.4	6.4	11.0	7.3
32	9.0	6.7	8.6	6.5	8.5	5.8	8.2	6.2
40	8.9	5.4	6.5	4.9	8.5	5.0	6.2	4.9
Mean	11.2	7.8	10.8	8.2	10.7	6.6	10.2	8.0
L.S.D. 5%	Treat. = 1.25 PXT = 2.8				Treat. = 1.14 PXT = 2.55			
C.V.	20.8%				20.2%			

TABLE 9

EFFECT OF DEBRANCHING AND DETOPPING TREATMENTS ON THE NUMBER OF PODS AND POD BEARING NODES ON MAIN STEM/PLANT OF SYRIAN LOCAL LARGE FABA BEAN

Population per m^2	Pods on main stem/plant				Pod bearing nodes on main stem/plant			
	T_1	T_2	T_3	T_4	T_1	T_2	T_3	T_4
8	5.3	9.1	6.8	3.1	5.1	7.5	6.1	2.6
16	4.1	10.1	4.7	3.1	4.0	8.4	4.5	3.0
24	3.3	7.5	4.7	2.7	3.3	6.4	4.4	2.4
32	4.2	6.7	3.9	3.2	4.1	5.8	3.8	2.9
40	3.6	5.4	3.2	2.6	3.4	5.0	3.1	2.6
Mean	4.1	7.7	4.7	2.9	4.0	6.6	4.4	2.7
L.S.D. 5%	Treat. = 0.63 TXP = 1.42				Treat. = 0.5 TXP = 1.14			
C.V.	20.5%				18.2%			

yield. Debranching (T_2) tended to increase the number of pod bearing nodes on the main stem and also resulted in a higher number of pods on the main stem (Table 9), but this did not compensate for the loss that occurred due to debranching.

It is evident from the foregoing that for drier environments of the type at Tel Hadia, branching is an important attribute and at least 3 branches/plant should be present to permit adequate crop canopy development and build-up of yield. Branching seems to be of special significance in relation to recovery from frost damage, since much of the recovery seems to occur by development of tillers from the lower nodes after the early foliage and shoot is killed. Hudeiba-72 and Giza lines had very sparse branching and these showed a higher kill due to frost (Table 2). The simulation of determinate growth habit by detopping does not seem to confer any advantage. Even the existing determinate mutant derivatives under our drier conditions tend to produce too little growth and form fewer reproductive nodes than seem desirable for optimum yield. The risk of late frost may be another factor limiting the usefulness of determinate growth habit under these conditions, in that there is likely to be little compensation for flowers lost due to frost.

OTHER FACTORS AND SUITABILITY FOR DRY AREAS

Another attribute of significance in the region is the high level of autofertility of the locally well-adapted genotypes. Data collected in 1980, in a flower tripping trial at Tel Hadia in an insect proof cage, revealed marked differences. Of the 36 genotypes evaluated, 8 were of diverse geographic origin and from open pollinated stocks and are shown in Table 10. There was clearly no need for tripping in Syrian Local Small, Giza-3, Giza-4, Hudeiba-72 and Seville Giant in contrast to Throws and Maris Bead, both of which showed very little autofertility.

TABLE 10

EFFECT OF TRIPPING ON THE POD AND SEED YIELD OF SOME SELECTED GENOTYPES OF FABA BEANS GROWN IN THE CAGE WITHOUT POLLINATING INSECTS

Genotypes	Pods/plant		Seed yield (g)/plant	
	Tripped	Untripped	Tripped	Untripped
Syrian Local Small	6.4	8.9	4.76	7.57
Giza-3	9.0	8.8	10.43	12.00
Giza-4	7.8	8.8	10.30	11.04
Hudeiba-72	5.6	7.1	4.88	6.56
Seville Giant	3.5	3.8	17.70	16.79
Express	5.1	5.0	16.11	9.74
Throws	3.6	0.8	3.65	0.46
Maris Bead	3.2	1.9	2.64	0.87

In addition to the above considerations, the plant ideotypes for drier areas need other desirable characters necessary for stable and high yields and acceptable seed quality. In this connection, resistance or tolerance to common diseases and pests, including the parasitic weed *Orobanche*, is worth mentioning. The early planted crop is particularly prone to attack by *Orobanche* and any degree of tolerance to this parasite in the genotype would make it particularly suitable for drier areas where early planting is desirable.

In summary, there are indications that faba bean plant ideotypes suitable for dry areas of West Asia and North Africa require the following attributes:

(i) Rapid early growth during winter.

(ii) Tolerance to frost during the early growth and ability to recover from the frost damage.

(iii) Ability to make rapid reproductive growth during the spring, so that most of the reproductive growth and yield development is completed before the environment becomes extremely hot and dry.

(iv) Moderate branching, 3 - 5 branches/plant, with a LAI of around 5.5.

(v) Ability to develop an elaborate and deep root system during winter and ability to extract moisture from deeper soil layers.

(vi) High autofertility.

(vii) Resistance to common disease and pests, particularly *Orobanche*.

(viii) Acceptable seed quality for human consumption.

REFERENCE

Harris, Hazel C., 1979. Some aspects of the agroclimatology of West Asia and North Africa. In: Food Legume Improvement and Development; Proceedings of a workshop held at the University of Aleppo, Aleppo, Syria, 2 - 7 May 1978. Ottawa, Ont. IDRC, 7-4.

R. Thompson *(UK)*

Perhaps we could open the discussion with Dr. Nagl's paper. To refresh your memories it dealt with comparisons between the conventional genotype and a range of terminal mutants in relation to components of yield including how the components of yield may compensate for each other; there were some new ideas of the potential of the terminal inflorescence type, which were very exciting in prospect.

D.A. Bond *(UK)*

Dr. Nagl said something about the height of pods above the ground in his mutants reducing apical dominance, would he elaborate?

K. Nagl *(Austria)*

The original determinate mutant with terminal flowers has the lowest pods too near the ground. However, in other genotypes pods are higher on the stem but, even so, the pods are perhaps too low.

D.A. Bond

On our trip yesterday to the polders we heard that the pods should be 30 cm above the ground for mechanical harvesting. Are some of yours as high as 30 cm?

K. Nagl

They are less than 30 cm but I do not know whether 30 cm is the right height anyway. I think that between 15 and 30 cm should be sufficient.

G. Dantuma *(Netherlands)*

I would like to comment on the 30 cm. This relates to long pods that hang down when they are ripe. The type of pod that Professor Nagl has is much shorter and therefore 15 - 20 cm in height above the ground would be enough.

R. Thompson

You mentioned that irrigation was a treatment which you are including this year. Were the plants you described today irrigated?

K. Nagl

All of the plants were irrigated because we wanted a non-competitive situation, therefore last year plants were wide-spaced and irrigated. This year we have both close and wide-spaced plants with and without irrigation to determine the response of the different genotypes.

D.A. Lawes *(UK)*

In his data, Professor Nagl showed that number of seeds per pod is reduced for the determinant types compared with conventional ones. Is the number of ovules per pod also reduced or only final seed number? Also, were there differences between the 23 lines for this character or were they all rather low?

K. Nagl

The values referred to the number of seeds. However, variation in this component was such that the selections for high numbers of grains/pod included some single lines which did not reach the level for this character of the initial determinate line. Nevertheless, selection for high grain numbers per pod resulted in a marked positive response in yield.

G. Pommer *(FRG)*

I would like to question the need for branching in various types of *Vicia faba*. We have heard that branching may be good in topless types and also in types with a short growth habit. In Europe we have types which tend to be tall. Can you comment on the interactions between branching, light penetration and yield? I would suggest that branching types and tall types may differ in terms of the most efficient harvest index because this depends on canopy growth and development. Is there any data on this?

K. Nagl

This particular aspect was not investigated but early development of vigorous branches, if development is simultaneous, would contribute to the yield. Even if the development of branches was only slightly uneven, irrespective of whether the plants were tall or short, the contribution would be reduced.

R. Thompson

Hopefully we shall get some answers on this question in the experiments you are doing this year on irrigation and density. It is a very interesting question as whether many or few branches are desirable and crops up time and again. As far as the density/yield relationship is concerned, branches and main stems are almost synonymous and the total number of fertile stems per unit area is the important factor, leaving aside considerations of uniformity of seed maturity. The question becomes more important when one considers whether large or small seeds are to be used. With the large seeded varieties there are obviously distinct advantages in having several branches because seed rates may be reduced.

R.B. Austin

With many crops, if they grow rapidly and are subject to

frost, they are found to be frost susceptible. It is clear that Dr. Saxena wanted to obtain fast growth in the winter so that he could achieve a measure of drought avoidance. Would there not be a conflict here between frost resistance and drought avoidance?

M.C. Saxena *(Syria)*

Yes, it would appear to be a conflict but basal branches must be taken into consideration. In spite of the fact that some of the genotypes had a very fast rate of growth, our observations on recovery last season showed that as long as the plants were branched there was considerable compensation. They would compensate to damage much earlier in development, and perhaps this was because the branches did not account for much assimilate; leaf area development was as great as would have been present without any frost damage. A fast rate of growth for a plant with a single shoot could result in 100% kill, which is what happened in 1972. But in other cases where there were three or four shoots already present, very fast new growth occurred and by the end of March there was little difference between the frost affected and non-affected plants.

R. Thompson

Thank you all for your contributions.

DETERMINATE GROWTH IN *Vicia faba*: AN OPPORTUNITY FOR ACCELERATED GENETIC TURNOVER

G.P. Chapman
Wye College, University of London,
Nr. Ashford, Kent, UK.

ABSTRACT

A combination of terminal and precocious flowering in Vicia faba *leads to a plant having relatively short stature, diminished canopy and appreciably shortened generation time. Such a plant can be sown at higher densities than has been customary for this crop and, utilising winter glasshouse facilities, two and even three generations per year seem possible.*

Such developments suggest analogies with cereals, particularly for example spring barley. This paper explores how far ideas derived from cereal breeding could be applied to Vicia faba, *and draws attention to the fact that variation in this species is sufficient to permit 're-modelling' to examine various agronomic problems.*

INTRODUCTION

Since its introduction by Sjödin (1971) the terminal inflorescence mutant ti has attracted attention and interest. Probably, it has more value not in converting existing indeterminate varieties but in a wider strategy that includes a re-appraisal of growth habit, planting density and time to maturity, for example.

The present author, in conjunction with colleagues, developed the concept of neoteny for *Vicia faba* and this is described in two earlier papers (Chapman, 1977; Chapman and Peat, 1978). Features of the neotenous plant include the following:

1. The vegetative phase is reduced substantially.
2. A reproductive apex replaces a vegetative one.
3. The last formed pods tend to mature first by reason of a favoured apical position.
4. Outgrowth of lateral buds to form 'tillers' is increased.
5. The chlorophyll-laden pod is placed above the canopy.

Such a plant offers a marked contrast to the standard leafy indeterminate growth habit and is consequently of interest to agronomists and physiologists. This plant type has been treated at Wye in two ways. It has been bulked to explore various aspects of its agronomy and physiology and it has been incorporated into a breeding programme to produce a range of sub-types modified in regard to particular characters.

REQUIREMENTS OF A MODEL

Remodelling a crop might be necessary because a wild or domesticated plant has inherent disadvantages, or because changing cultural practices require a modified plant. A crop model is an actual plant utilised experimentally. It represents an approximation to some ideotype. Three features sought in any crop model should be that:

1. Sinks comprising the economic yield accumulate their contents with improved efficiency.
2. New agronomic options are added to those already available.
3. The new model is not less and preferably more susceptible to genetic improvement than the type it replaces.

When confronted with a wide range of variants from which to choose, two problems arise, namely which variants to select for bulking and under what kind of regime should they be tested.

NI* UNDER FIELD CONDITIONS

Bulking of seed has been in progress for two years and during 1980 it should be possible to obtain the first realistic measure of performance. From small plot trials the following have emerged:

1. Density

Plants at densities up to 250 per square metre have been examined. At high density, branch number is reduced but the proportion of fertile branches increases. At all densities there was relatively little lodging.

2. Crop cycle

During 1979 the crop cycle was completed in less than 120 days at Wye. Since 1980 was a wetter year, harvesting will be later, however, than for 1979, though earlier than for indeterminate plant types.

* See later section 'Modification of Neoteny'.

3. Winter performance

Most autumn sown plants survived what was a relatively mild winter and conclusions about hardiness would be premature. Plants were shorter than those spring-sown.

4. Time of harvesting

During 1980 time to maturity is being studied relative to Goldspear spring barley. The two crops were sown simultaneously on 8th April and setting of the terminal inflorescences coincided with heading-out of the first cereal plants on June 10th.

5. Pod set in relation to aphid migration

Plants sown on 26th March showed a very high proportion of set pods by 15th June, 1980. There is the possibility with this kind of plant model that the most vulnerable stages of plant growth will precede the main part of the aphid population growth (Apterae → Alatae on summer hosts). Aphid records for previous seasons are at present being examined.

MODIFICATION OF NEOTENY

While the plant just described represents an interesting model, it is unlikely to be the only or the best such that could be produced. A crossing programme was devised therefore, to examine how this type might be modified.

The first plant type thus produced is referred to here as NI, and II, III and IV its subsequent, more or less parallel, derivatives. The three newer types were recombined to give F_1s and F_2s and the individual plants in the segregating generations selfed on one inflorescence.

Some 1 200 F_2 plants were set out and divided among three planting dates, the first to search for maximum yield, and the last two to detect if possible evidence of aseasonality. The population showed considerable diversity in height, vigour, leafiness, branching, earliness and likely yield. Additionally, at maturity differences in seed size, colour, shape and number per pod will be scored.

What is already apparent is that the three sub-types NII, III and IV have produced an array of intergraded forms capable of being used as models under particular agronomic or physiological conditions.

GENETIC TURNOVER

The shortened generation time seems likely to provide the opportunity for an F_2 to be tested in consecutive years under field conditions while permitting the parental and F_1 generations to be raised under glass in winter*.

Wye College in conjunction with the Rank Hovis McDougall Company, is currently testing such a procedure.

Obviously, the scheme could be applied to alternative patterns of three-generation years. One example would be the use of trisomic stocks, the development of which was begun by Martin (1978) for accelerated linkage mapping, when the complete set is available.

CONCLUDING COMMENTS

Neotenous plant models described here show relative freedom from lodging, some prospect of aphid avoidance, earliness of harvest, harvestability and some possibility of accelerated genetic turnover. Perhaps, therefore, a derivative could fit fairly readily into present day, cereal-dominated agriculture.

The indeterminate field bean grown in several years and locations shows, within the same general morphology, a wide range of expression and yield performance. Yield might sometimes be high but it is very variable.

* See Appendix.

By contrast growth of cereals is both simple and strongly determinate. It is, therefore, relevant to ask if the production of a neotenous *Vicia faba*, such as for example those described here, could offer relative consistency. To this extent there would be an analogy with cereals. If, however, a comparison is made with (say) spring barley, the following additional similarities exist:

i) There are relatively few chromosomes which, with a combination of cytological techniques are individually characterisable.

ii) Following the work of Martin (1978) there is the possibility of developing the expected set of trisomics.

iii) There is an array of single locus mutants available.

iv) Three generations per year almost certainly are possible, utilising appropriate winter glasshouse facilities.

v) Generation time under field conditions (and height at maturity) are similar.

While clearly the degree of self-pollination in barley is much higher, the rate of progress possible in *Vicia faba* might, with an appreciable expansion of effort, now begin to match that possible for a major cereal.

REFERENCES

Chapman, G.P., 1977. Re-structuring the field bean (*Vicia faba* L). Scot. Hort. Res. Inst. Bull. 15, 3-9.

Chapman, G.P. and Peat, W.E., 1978. Procurement of yield in field and broad beans. Outlook on Agriculture 9, 267-272.

Martin, A., 1978. Aneuploidy in *Vicia faba*. Jour. Hered. 69, 421-423.

Sjödin, J., 1971. Induced morphological variation in *Vicia faba* L. Hereditas 67, 155-180.

APPENDIX

WINTER GLASSHOUSE ARRANGEMENTS

For 12 hours the minimum night temperature was 10°C and then heating was increased to give maximum day temperatures of 22°C. Fluorescent tubes or high pressure sodium lamps were used to provide 16 hours light. The heat output from the lamps in the last four hours of each day moderated the rate of cooling.

RELATIONSHIP BETWEEN PLANT DENSITY AND STRUCTURE OF YIELD IN DIFFERENT GROWTH TYPES OF *Vicia faba* L.

E.R. Keller and J. Burkhard

Institute of Crop Science,
Swiss Federal Institute of Technology (ETH),
Zürich, Switzerland.

ABSTRACT

Based on previous results showing that the harvest index of Vicia faba *was relatively stable, it was concluded that grain yields could be improved by increasing the total dry matter production per unit area. Information was required on which to base the selection of plant densities for different types of* Vicia faba *for use later in larger field experiments. With this purpose in mind, a field experiment was conducted in 1979 using a Nelder systematic spacing design with continuously increasing distances between the rows and between plants within each row. The results showed clear differences in the zone of pod set on the main stem, varying with density and genotype (Herz Freya, Minica, Maris Bead, Svalöf 0621). It was apparent that grain yield and total dry matter production both increased as density was increased from 10 plants/m^2 up to 80 plants/m^2. The harvest index decreased slightly with increasing density; there were, however, varietal differences in this respect as well as in regard to the influence of density on some yield components.*

INTRODUCTION

In their experiments with the field bean variety Herz Freya (*Vicia faba* L.), Gehriger (1978) and Bellucci (1980) came to the conclusion that the harvest index (HI) is, in general, a stable factor. It can be seen (Table 1) that topping the plants did not influence the harvest index. The figures presented in Table 2 show the influence of gibberellic acid, applied at different concentrations. Here again, the harvest index was not influenced significantly. It is remarkable that in all experiments carried out at our Research Station at Eschikon with the variety Herz Freya during 1976 - 1979, the harvest index did not change very much from year to year. This stability was rather surprising, so we conducted 13 additional field experiments in 1979 at different locations. It may be concluded that the HI (Table 3) does not vary very much from location to location with the exception of Nyon and Goumöens-la-ville. The trial at Nyon (Federal Research Station) suffered from serious drought, and that at Goumöens-la-ville was severely damaged by aphids. The HI for the variety Herz Freya, grown under normal conditions but at different locations and with different yield/ha, proved again to remain relatively stable. On the other hand, Gehriger (1978) demonstrated a close relationship between leaf area duration (LAD) and grain yield (Figure 1). We therefore concluded that grain yields may be improved by increasing the total dry matter production per unit area i.e. a factor which depends much on growth type and plant density.

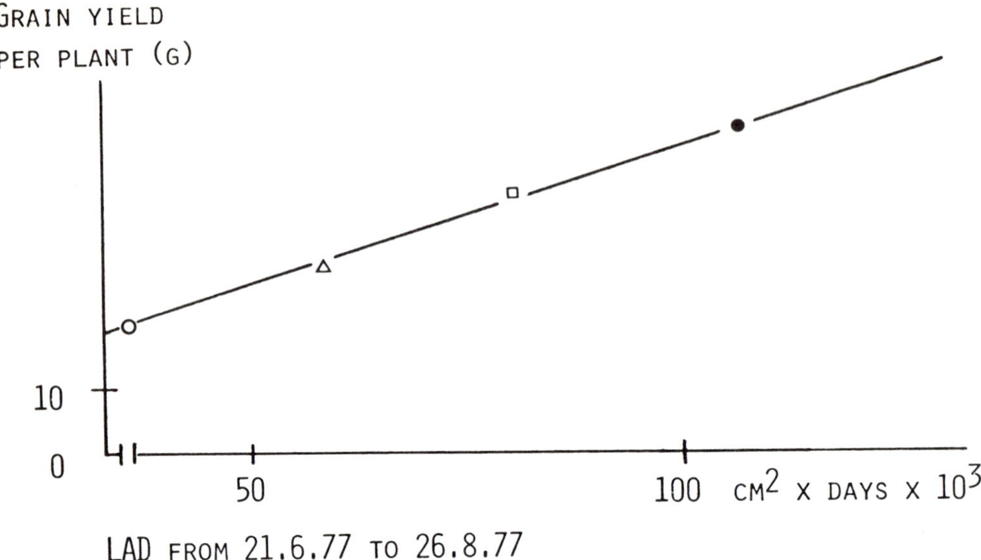

Fig. 1. Relation between leaf area duration (LAD) and grain yield (Gehriger, 1978).

o: T, △: T_2, □: T_3 (see Table 1).

FIELD EXPERIMENT WITH DIFFERENT VARIETIES i.e. GROWTH TYPES AND PLANT DENSITIES 1979

Experimental design

In order to compare a wide range of plant densities we used a Nelder systematic spacing design (Nelder, 1962) with continuously increasing distances between rows and between plants within each row (Figure 2). This design required the germination and growth of all seeds (i.e. no gaps); we therefore planted watersoaked and pregerminated seed.

The varieties selected to represent different growth types were Herz Freya, Maris Bead and Svalöf 0621 (topless). Herz Freya normally produces one stem and the others several per plant. The growth conditions in 1979 were good.

TABLE 1

HARVEST-INDEX AS INFLUENCED BY TOPPING AT DIFFERENT DATES (GEHRIGER, 1980). VARIETY - HERZ FREYA

Time topped after flowering	Days	1976 HI	1976 Grain yield rel.	1977 HI	1977 Grain yield rel.
Control		0.57	100	0.54	100
T_1	9			0.55	54
T_1	4	0.54	78		
T_2	13			0.53	63
T_2	70	0.51	84		
T_3	20			0.54	85
T_3	10	0.54	77		
Significance		NS	NS	NS	XXX

Harvest index: $\frac{\text{Dry matter grain/plant}}{\text{Total dry matter/plant}}$

NB. (Calculated 1976 without root weight)

TABLE 2

HARVEST-INDEX, INFLUENCED BY THE APPLICATION OF GA_3 AT DIFFERENT CONCENTRATIONS AND DATES (BELLUCCI, 1980). VARIETY HERZ FREYA.

Treatment	1978 HI	1978 Grain yield rel.	1979 HI	1979 Grain yield rel.
Control	0.48	100	0.56	100
GA_3 (4 leaves, 3 x 10^{-4}m)	0.49	96		
GA_3 (4 leaves, 10^{-4}m)	0.50	128		
GA_3 (4 leaves, 3.3 x 10^{-5}m)	0.47	85		
GA_3 (6 leaves, 3 x 10^{-4}m)	0.49	87		
GA_3 (6 leaves, 10^{-4}m)	0.49	140		
GA_3 (6 leaves, 3.3 x 10^{-5}m)	0.49	107		
GA_3 (5.7 leaves, 10^{-4}m)			0.56	108
GA_3 (6.5 leaves, 10^{-4}m)			0.57	116
GA_3 (7.4 leaves, 10^{-4}m)			0.57	103
GA_3 (10 leaves, 3 x 10^{-4}m)	0.47	89		
GA_3 (10 leaves, 10^{-4}m)	0.49	122		
GA_3 (10 leaves, 3.3 x 10^{-5}m)	0.48	87		
Significance	NS	XXX	NS	NS

TABLE 3

HARVEST-INDEX AS INFLUENCED BY LOCATION 1979. PLANT DENSITY: 33 PLANTS/m^2 (BURKHARD, 1980). VARIETY HERZ FREYA

Location	HI	Grain yield q/ha	Remarks
Nyon	0.37	23.1	Drought
Goumöens-la-ville	0.26	13.7	Aphids
Payerne	0.50	47.8	
Courgenay	0.44	43.3	Low density
Langnau i.E.	0.57	28.4	(10 - 12 plants/m^2)
Langenthal	0.47	50.3	
Wohlenswil	0.50	50.8	
Eschenbach	0.54	45.9	
Herrliberg	0.47	38.7	
Eschikon-ETH	0.51	51.0	
Klingenberg	0.57	46.3	
Amlikon	0.60	49.2	
Malans	0.53	35.3	

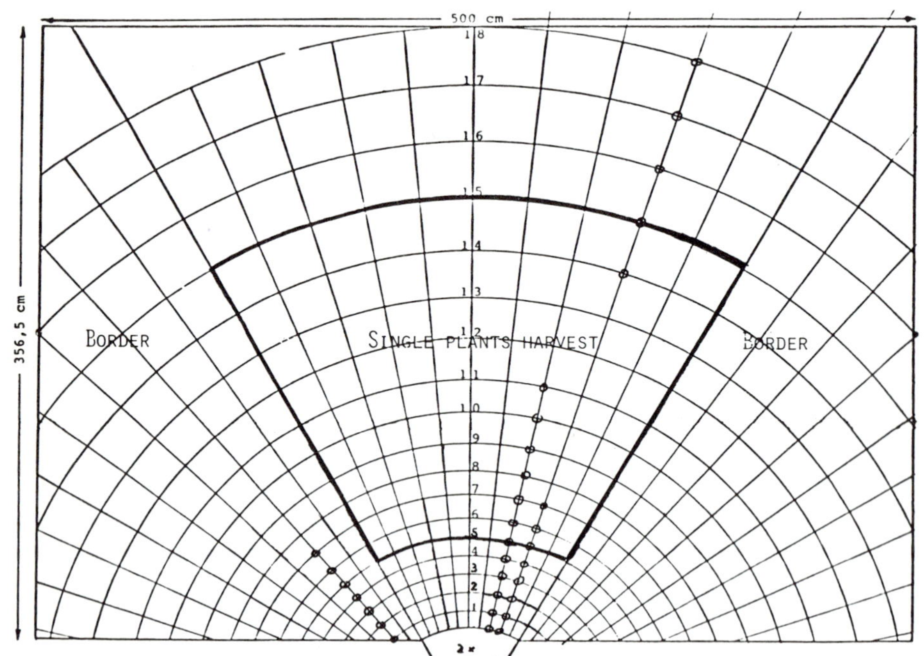

Fig. 2 Nelder-design of an experimental spacing plot. 387 single plants = 0. Border effects eliminated by guards.

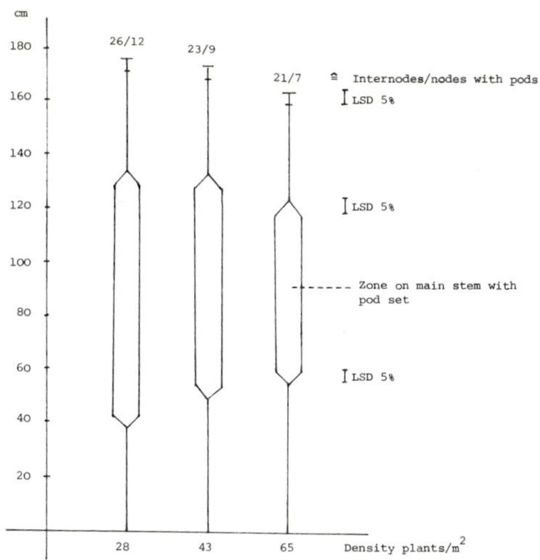

Fig. 3 Plant length, zone of pod set and number of internodes at three selected densities. Nelder-design, 1979. Variety: Herz Freya. (Mean value from 44 plants).

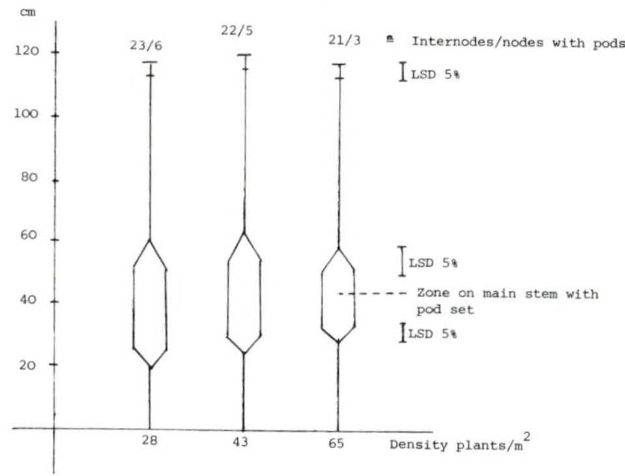

Fig. 4 Plant length, zone of pod set and number of internodes for three selected densities. Nelder-design, 1979. Variety: Minica. (Mean values from 44 plants).

Results and discussion

1. Influence of plant density on the growth type of the different varieties. The zones of the pod set on the main stem are given in Figures 3 - 6 for three different plant densities. The height of the lowest pod was greater the higher the plant density, and the number of nodes over which pods were borne tended to be reduced at the higher densities. Only a small portion of the stem of Minica and Svalöf 0621 was utilised for the pod set. Could this mean, in the case of Minica, that the yield potential was not fully utilised?

2. Influence of plant density on yield of different varieties. From Figure 7 (log scale), it is apparent that grain yield per unit area increased up to a density of 80 plants/m^2, but only if lodging was absent; lodging was prevented in this trial. Since the calculation of the yield is based on 44 plants for each density, the yield level is probably greater than in the field. Of special interest is the continuous increase in the yield of Maris Bead, associated with increasing density.

3. Influence of plant density on the harvest index of different varieties. As expected, the harvest index decreased with increasing plant density, due in part to less favourable light conditions (Figure 8). The differences, however, were relatively small and the varieties probably do not behave in the same way. The remarkably slight change in the HI of Maris Bead, between densities of 10 and 80 plants/m^2, helps to explain the increase in grain yield along with the marked increase in total dry matter production associated with increasing density (Figure 9).

The results from this trial confirm that the production of increased dry matter per unit area is, under the prevailing conditions, correlated with an increase in yield. That such a correlation seems to be valid, even for a very high density, is rather surprising.

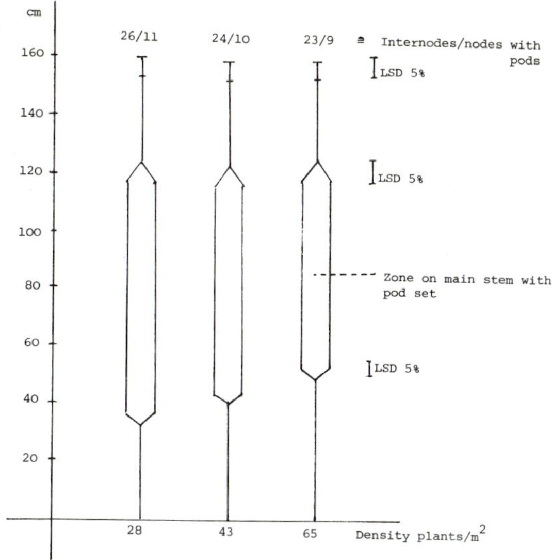

Fig. 5 Plant length, zone of pod set and number of internodes at three selected densities. Nelder-design, 1979. Variety: Maris Bead. (Mean values from 44 plants).

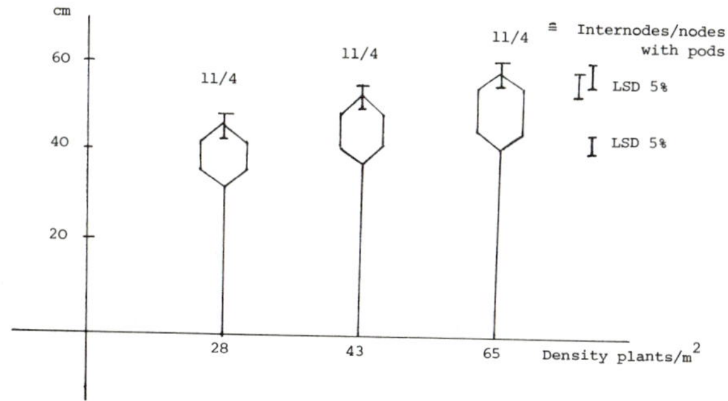

Fig. 6 Plant length, zone of pod set and number of internodes at three selected densities. Nelder-design, 1979. Variety: Svälof 0621. (Mean values from 44 plants).

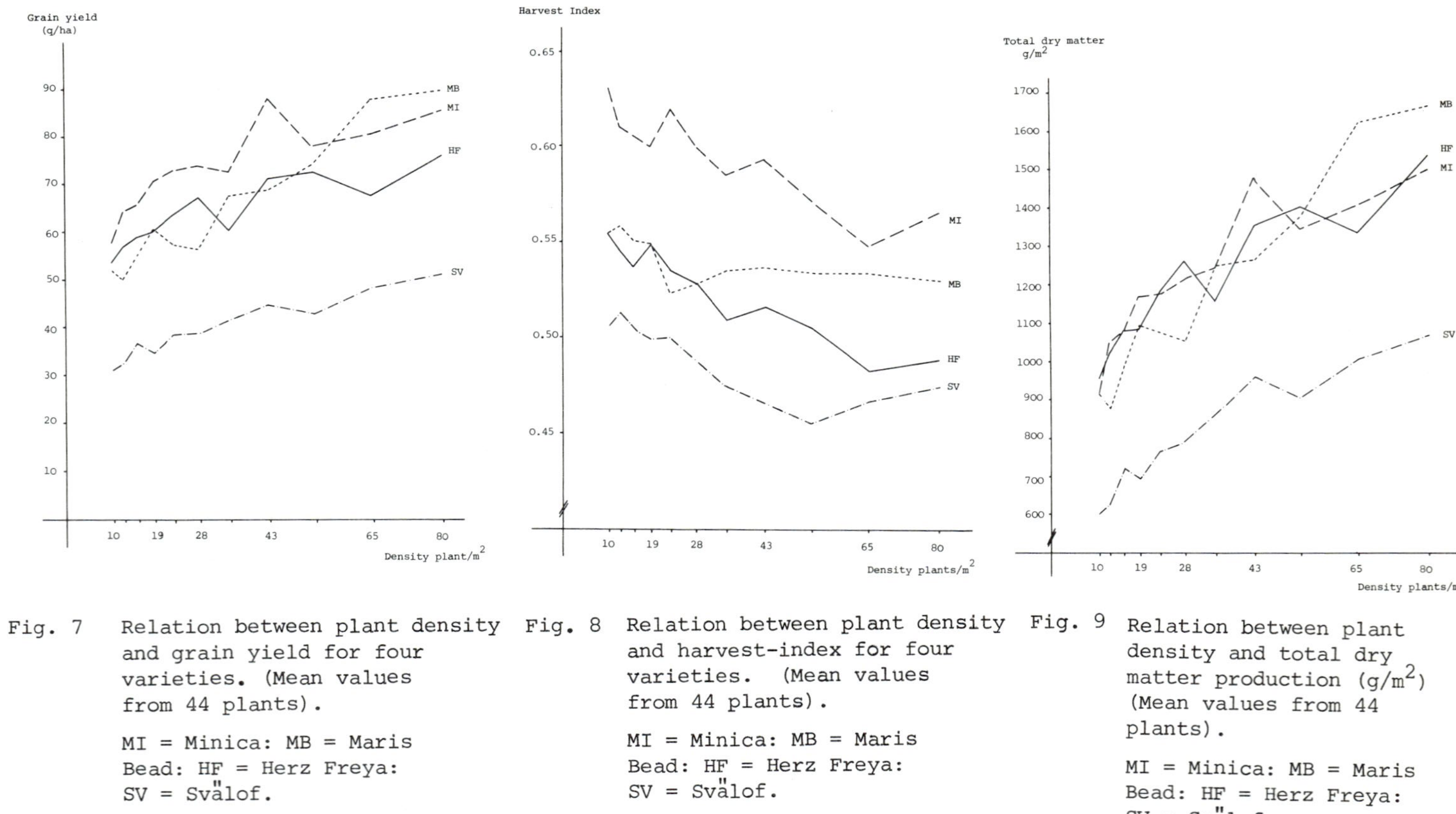

Fig. 7 Relation between plant density and grain yield for four varieties. (Mean values from 44 plants).

MI = Minica: MB = Maris Bead: HF = Herz Freya: SV = Svälof.

Fig. 8 Relation between plant density and harvest-index for four varieties. (Mean values from 44 plants).

MI = Minica: MB = Maris Bead: HF = Herz Freya: SV = Svälof.

Fig. 9 Relation between plant density and total dry matter production (g/m^2) (Mean values from 44 plants).

MI = Minica: MB = Maris Bead: HF = Herz Freya: SV = Svälof.

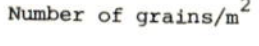

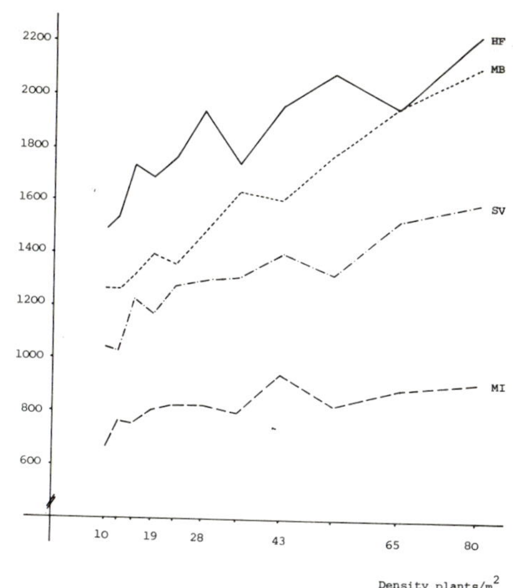

Fig. 10 Relation between plant density and number of produced grains/m^2 for four varieties. (Mean values from 44 plants).

MI = Minica: MB = Maris Bead: HF = Herz Freya: SV = Svälof.

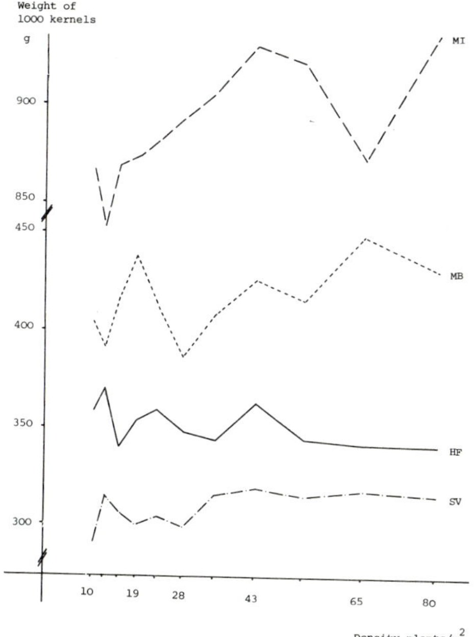

Fig. 11 Relation between plant density and 1000 seed weight for four varieties. (Mean values from 44 plants)

MI = Minica: MB = Maris Bead: HF = Herz Freya: SV = Svälof.

4. Influence of plant density on some yield components of different varieties. Although the number of pods and seeds/plant decreased rapidly with increasing density, the total number of pods and grains per unit area increased (Figure 10). There were some differences between the four varieties in this respect reflecting the different growth types. Minica showed only a slight increase in grain number, whereas the 1000-kernel weight increased much more with increasing density than for other varieties (Figure 11).

FINAL CONSIDERATIONS

The systematic spacing experiment was laid out in 4 replications in one year. The results, therefore, have to be interpreted with some caution. One of the aims of the experiment was to gain information on the selection of plant densities for the larger field experiments in which the behaviour of yield components, with and without the application of GA_3, is to be analysed. Based on the systematic spacing experiment, we established two field experiments in 1980 at different locations with the varieties Herz Freya and Maris Bead, and with densities of 30, 40, 50 and 70 plants/m^2.

The possibility that yields of all varieties used in the systematic spacing experiment were still increasing with very high plant densitites led to the conclusion that resistance to lodging respresents a high priority breeding objective. On the other hand, there seems to be little difference between varieties with one or more stems/plant in their capacity to increase total dry matter production per unit area as density is increased. This is probably due to compensation in stem numbers in respect to increased competition at higher densities. Finally, we believe that the varietal differences in the zone of pod set indicate to the breeders ways in which to increase the harvest index still further.

REFERENCES

Burkhard, J., 1980. Interner Bericht. Institut für Pflanzenbau ETH, Zürich.

Gehriger, W., 1978. Influence de la temperature et de l'écimage sur le developpment de la féverole (*Vicia faba* L.) et étude de la nutrition des fleurs en assimilats marqués au 14 C. Diss. ETH, Zürich.

Jacquiéry, R., 1977. Etude de la chute des fruits chez la féverole (*Vicia faba* L.) relations avec la disponibilité en assimilats marqués au 14 C. Diss. ETH, Zürich.

Nelder, J.A., 1962. New kinds of systematic designs for spacing experiments. Biometrics, 18, 283-301.

DISCUSSION

G. Pommer *(FRG)*

It would be interesting to hear from Dr. Chapman on yield relationship between his types and the normal types of *Vicia faba*.

G.P. Chapman *(UK)*

In my paper I pointed out that we have a field trial this year. Other than that I think it is improper and illadvised to speculate on acreage yields from small plots. I am as anxious as you are to get some answers to that question.

M.H. Poulsen *(Denmark)*

I would like to ask Dr. Chapman about the conditions under which he can grow three generations per year and get good seeds which germinate well. I am thinking in particular about details of the environment in the glasshouses.

G.P. Chapman

The glasshouse was used routinely for producing barley and it was taken over without modification for producing field beans. There was supplementary lighting and heat. It was really quite straightforward to arrange and it did not present any great problems.

G. Duc *(France)*

I would like to ask Dr. Keller about his type of measurement of harvest index. Is the weight of the roots always considered in the measurement of harvest index (HI)?

E.R. Keller *(Switzerland)*

It was not possible to include root weight in all measurements. As indicated in my Table, in some, roots were included but those in the Nelder systematic design experiments did not include the roots. It could make a difference in the harvest index of 5 - 10%.

J. Picard *(France)*

How was harvest index estimated in the different situations? Is it based on separation of the grain and the other parts of the plant at harvest time, or on the maximum above ground dry matter yield achieved in relation to final seed yield?

R. Thompson *(UK)*

This is something which needs clarifying. HI is a widely used term and originated in relation to cereals where all the plant components, including leaves, remain at maturity, which makes its calculation unambiguous. But with beans, leaves fall off, and can make comparisons rather difficult. One way around this with some experiments is to plot the development of harvest index with time, which incidentally provides for additional physiological understanding. Perhaps we should agree on what we mean by harvest index with field beans.

G. Dantuma *(Netherlands)*

There are several methods, but I think we have agreed that this year, for the joint field bean tests, it is to be simply the ratio of seeds to the rest of the plant at the time of harvest; that is the only practical way to do it - not the total biomass.

M.H. Poulsen *(Denmark)*

I think that what Dr. Dantuma has said is a very practical and sensible solution; we have done the same. In addition, we have tried to separate the pods into shells and seeds to find genotypes with a low shell/seed ratio. This is possible but it takes longer.

R. Thompson

Before we move on I would like to comment on Professor Keller's paper. Here again, we have a situation in which beans of a number of different genotypes are responding differently to plant population. From the data presented at this conference it appears that there is a lot of information available in terms of the response of different genotypes to different environments. The big problem is to collate all of this and perhaps group meetings of this sort should be extended to include this objective.

G.P. Chapman

Paul Hebblethwaite (of Nottingham University) is editing what is presumed to be the definitive work on *Vicia faba*, along with other people and I think that is the appropriate focus for the sort of information you have been talking about.

R. Thompson

Thank you, Dr. Chapman and Professor Keller; we must move on now to the next paper.

SURVEY OF THE BREEDING WORK ON *Vicia faba* AT VEG SAATZUCHT GOTHA/FRIEDRICHSWERTH*

M.H. Poulsen
Danish Plant Breeding Ltd.,
Boelshøj, 4660 Store-Heddinge, Denmark.

ABSTRACT

VEG Saatzucht Gotha/Friedrichswerth is the centre for practical breeding work on Vicia faba *in the German Democratic Republic. It comprises two stations, one in Gotha and one in Friedrichswerth, some 20 km apart. Dr. agro Habil Rudolf Steuckardt is the leader of the breeding work on which 22 scientifically trained co-workers, plus a number of technical assistants, are engaged.*

The work on Vicia faba *is divided into basic research work, practical breeding and maintenance breeding. The basic research work comprised an eight year programme on comparisons among the effects of mass selection, poly-cross and top-cross. Further, the relationships between nitrogen content, seed yield and protein quality were studied. Studies on cross pollination had shown the outcrossing to be approximately 50%.*

Work on mutation, disease resistance, autofertility, F_1 hybrids and yield physiology are in progress.

The practical breeding work comprised a large area of experimental/evaluation plots. The pedigree method is used extensively. Several new types were being or had been examined, but there was still much faith in the conventional type as represented by the two German Democratic Republic varieties Fribo and Erfordia.

* Report of a visit on 24th and 25th of July, 1979.

INTRODUCTION

The field bean acreage in the German Democratic Republic is about 25 000 ha and the average yield 3.5 t/ha.

For successful cropping, the following aspects of management are considered to be important:

1. As early sowing as possible.
2. Deep sowing, 8 - 12 cm.
3. 40 - 50 plants/m^2.
4. High soil fertility; application of 30 kg N/ha as a starter.
5. Effective chemicals, weed and pest control.
6. Honeybees for pollination, four colonies/ha. (The fields are generally not less than 100 ha).
7. Spraying with a desiccant (reglone) before harvest.
8. Gentle threshing and drying of the seeds.

VEG Saatzucht Gotha/Friedrichswerth is one of 100 farms scattered over the German Democratic Republic with the commitment to: 1) breed new varieties; 2) maintain varieties and 3) produce seeds for grain production on about 6 000 000 ha which is divided into production farms, each of approximately 5 000 ha. Saatzucht Gotha/Friedrichswerth has 2 000 ha and is located about 20 km south-east of the Harz at Gotha (Sundhausen), and Friedrichswerth.

Apart from field bean which is bred at both locations, winter wheat is bred at Gotha and winter barley at Friedrichswerth. Dr. Rudolf Steuckardt is the leader of the breeding work, with a team of 22 co-workers with university education, and a varying number of technical assistants are engaged.

Field bean breeding is organised by Dr. Steuckardt at Berthelsdorf and Gülzow, as well as at Gotha/Friedrichswerth.

These four breeding stations collaborate closely to exploit fully the breeding material as well as the facilities. There is, however, some specialisation in respect of the various breeding methods.

At Gotha work is carried out on breeding methodology. Here 5 years of a planned 8 year substantial programme have been completed. Comparisons are made of the effects of mass selection, poly-cross and top-cross on production of synthetic varieties.

RESEARCH

Mutation

Mutation breeding is in progress using ethylmethylsulph-ate (EMS), sodium azide and X-radiation. Treatments were given recurrently with an intensity allowing 50% survival. The mutagen-treated material was screened for: 1) increase in the additive genetic variability of the common indeterminate type; 2) new terminal inflorescence types; 3) types with stable male sterility and 4) increased protein quality. Many examples of the various mutants described in the literature were found in the material. Additionally, a new so-called kleistogamous flower occurred which has a very compressed and 'extended bud stage' appearance. The ovary and style are pressed together by the still folded petals so that ovary and style lie in folds within the keel. The plants in the population which I saw varied from some completely sterile to a few plants with normal fertility, when judged from the number of seeds set. Some plants developed a few apparently normal flowers among the mutant flowers. The flower-type of the fully fertile plants could not be judged as flowering had terminated on these plants. Selection for increased fertility was planned. With this flowertype, complete protection against bee-visits is possible.

Resistance to diseases

Selection for resistance to *Ascochyta* is in progress.

Infestation of the breeding material is done by mixing the seeds just before sowing with a substrate (wheat grains) containing the fungi. This year the infection was successful. The offspring of the few healthy plants would be tested next year. No results on the effects of selection are yet available.

Seed quality

The chemical composition of the seeds is being studied with respect to protein content and quality. The current work revealed that among high yielding entries there existed a negative correlation between seed yield and protein content. Furthermore, the protein content was negatively related to the protein quality as expressed by the amount of sulphur containing amino acids and of lysine.

Physiology

At Friedrichswerth, work is carried out on yield physiology and morphology. This work is directed by Dr. H.F. Kästner.

The yield of the plant is determined by the four yield components: number of inflorescences, number of pods per inflorescence, number of seeds per pod and seed weight. The values for these components are thought to be determined successively - in the order given above - by the assimilatory capacity of the plant at the different stages of development.

Studies on translocation of photosyntates had shown that the direct effect of leaves on seed yield ceased when the seeds had a dry matter content of 40%. New experiments were planned to study the effect of desiccation of the leaves by application of reglone when the dry matter contents of the seeds were 20, 40, and 60%.

By application of gibberelinit was demonstrated that the common dwarf habit found in *Vicia faba* resulted from reduced gibberelin synthesis in the dwarf plant.

Various minor changes of the basic plant model had been suggested with the aim of increasing the harvest index. But the yield of all such 'new types' always fell below that of the traditional types Fribo and Erfordia, which have long, fairly thick stems and a heavy foliage. A terminal inflorescence mutant type, similar to the one from Svalöf, was tested. The new type appeared good but, in reply to a question about the yield of this type, my impression was that it would be about 20% lower than that of the normal type.

Inbreeding

Inbreeding by selfing was done in cheese cloth tunnel cages, approximately 1.5 m wide and 1.30 m tall. The length varied from 5 to 40 m. A few larger semi-cylindrical cages were also used. Samples of ten varieties had been inbred by selfing for eight generations. Many of these inbred plants had set a normal amount of pods and the plants did not appear to suffer from inbreeding depression with respect to fertility and vigour. Selection was to be made for auto-fertility and combining ability with the aim of producing synthetic varieties.

It had already been shown that compared with the basic population the yield of the inbreds decreased during the first four or five generations of selfing after which the yield appeared to stabilise despite further inbreeding. One generation with open-pollination was sufficient to restore the yield of the 'inbred' populations to the levels of the basic population.

Out-crossing

Studies on the frequency of out-crossing at various positions on the plant were made using the white flower and black seed coat characters. Frequencies of around 50% had been found at all positions. (Here it may be mentioned that four honeybee colonies/ha were provided in order to ensure adequate pollination). In this connection three rows of oats between the plots were tried as a guard against intercrossing between the plots. Preliminary results suggested that, by this method, crossing between plots could be reduced to approximately 15%.

The philosophy behind the choice of synthetic varieties is the wish to provide the farmers (for additive genetic reasons) with high yielding material which, in addition, may also show a level of yield heterosis related to the amount of crossing in the previous generation. It was agreed that no guarantee could be given for the effects of heterosis and that it probably would vary over generations. But it is the opinion here that the chance for heterosis should be exploited and also in new autofertile synthetics if such could be produced in the future.

Dr. Steuckardt explained that, at Berthelsdorf, additional work on autofertility was in progress. The autofertility of exotic types of field beans was transferred into the normal European types. At Gülzow studies were performed on the exploitation of male sterility in the production of F_1 hybrid varieties.

Practical breeding work

Apart from this division of the different aspects of breeding technique, all the stations were involved in the practical breeding programme. In this work the material was handled as if no contamination crossing between plots took place.

The practical breeding process can be described as follows:

1. Selection of single plants from new breeding populations or from later steps in the breeding programme. At Gothe/Friedrichswerth 11 000 plants were selected annually. They were designated A-lines and each was hand sown in a small plot (1.5 m^2) for visual evaluation.

2. One thousand, eight hundred of the A-line plots were selected and harvested as B-lines, to be sown the following

year by machine in 12 m^2 plots. These plots were harvested by combine and all which yielded at least 15% more than the control variety Fribo were selected as C-lines.

3. The C-lines were tested in trials with four replications (12 m^2 plots) and the 20 - 30 best C-lines formed characteristic I-lines.

4. I-lines were mixed and this new population was multiplied and tested by the breeders in different locations for two or three years.

5. Preliminary official trials followed at 12 locations for 1 - 2 years.

6. Official trials at 35 locations for 2 - 3 years.

Maintenance breeding was done by the same method. The A-lines were recurrently selected from the individual B and C-lines so that the characteristics of the I-lines (offspring of C-lines) were maintained. The variety Fribo is based on 23 I-lines which are mixed and multiplied for 3 - 4 generations, after which the farmers may use their own seeds for another 3 - 4 generations. It was argued that despite the open-pollination it was possible to maintain the characteristics of the I-lines and therefore the genetic composition of the variety. Simple bulk multiplication was thought to change the variety. A possible change of the variety due to the described maintenance selection has not been examined. The maintenance is not performed by the breeder of the variety. Thus Erfordia is maintained at Erfurt and Fribo at Leipzig.

All the breeding material which I saw appeared fairly uniform in type. The A-lines were somewhat distinct, but at later stages the variation appeared to be greater within lines than between lines.

SOME OBSERVATIONS ON HETEROSIS IN SUCCEEDING GENERATIONS OF SINGLE CROSSES OF *Vicia faba*

E. von Kittlitz
Landessaatzuchtanstalt, Universität Hohenheim,
D-7000 Stuttgart-70, Postfach 106,
Federal Republic of Germany.

ABSTRACT

When comparing F_2 *and* F_3 *progenies of 3 crosses between inbred lines to the parental means, no indication of heterosis in plant yield or yield components could be shown under open pollinating conditions (i.e. with insect pollination). Growing the same material in bee-proofed cages, we got heterotic effects as expected, being at a maximum in* F_1 *and gradually decreasing from* F_1 *to* F_3 *generations.*

The results are briefly discussed with respect to autofertility and possible consequences on breeding strategies of synthetic varieties.

INTRODUCTION AND METHODS

Preliminary studies have been made to examine yield potentials of synthetic populations of *Vicia faba* and initial examinations were particularly concerned with heterotic effects in F_1, F_2 and F_3 progenies of single crosses.

Seed for the experiment of the F_1 generation, the F_2 generation and one part of the F_3 (designated F_{3II}) were produced (Figure 1) in isolation cages, thus the F_1 and F_2 seed resulted only from selfing. The other part of the F_3 (designated F_{3I}) was from an isolated plot at some distance from other faba bean plants, and was partly cross- and partly self-pollinated. In the experiments the following progenies of 3 single crosses of 3 inbred lines (L1 x L3, L1 x L7 and L3 x L7) were studied: F_1, F_2, F_{3I}, F_{3II}, P_1 and P_2. The parent lines used were typical representatives of the sub-species *minor* (L3), *equina* (L7) and *major* (L1). The average proportion of heterozygous loci is given in Figure 1. In this respect, F_1, F_2 and F_{3II} as well as the parents need no further explanation. In the F_{3I} the following assumptions were made:

1. That in the F_2s 50% of the flowers set pods through self pollination.
2. That the other 50% of the flowers were open to a random distribution of cross and self pollen, and an equal chance for cross- or self-fertilisation.

This means that possible differences in autofertility between the progenies did not affect the degree of selfing when the insect population was able to ensure adequate pollination. Such relations are clarified (Figure 2) by examination of the offspring of the dominant homozygous F_2 plants, within the limits of the above assumptions. The offspring produced by selfing these plants are 100% dominant homozygous individuals and by open-pollination comprise 50% dominant homozygous and 50% heterozygous. In terms of the whole population 12.5% homozygous dominants are produced by selfing of the dominant genotype and

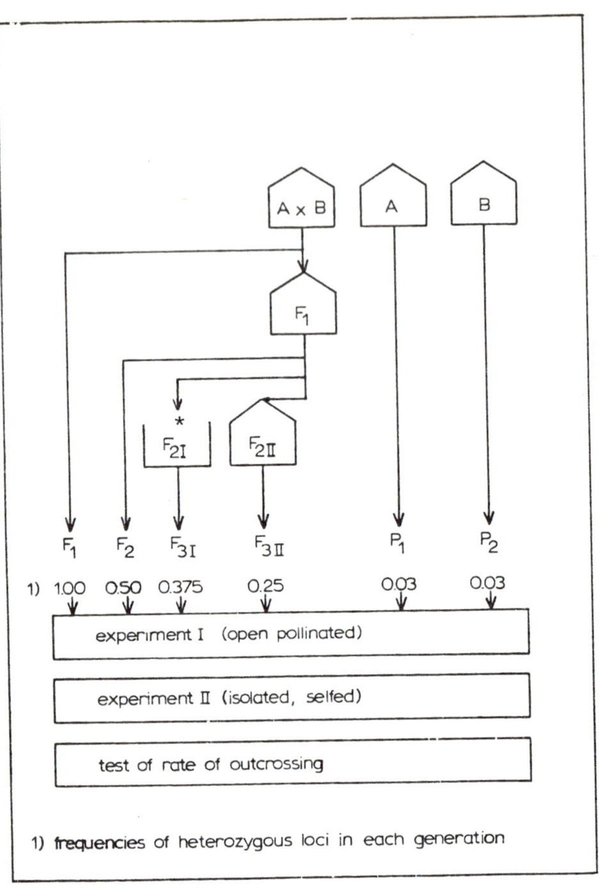

Fig. 1: Structure of the test generations
* open pollinated in an isolated plot

6.25% homozygous dominants and 6.25% heterozygous individuals are produced following open pollination. Summing up, there are 37.5% heterozygous loci in the population (Figure 2).

Results and discussion of two experiments on heterosis

In both experiments we used a simple, randomised block design, and all progenies were grown in 3 replicates each consisting of 10 plants. Experiment I was grown under open field conditions, i.e. with insect pollination. In Experiment II

Rate of selfing and outcrossing	S 0.50		C 0.50

F_2		F_3		
AA 0.25	1	(AA 1.00)	+	(AA 0.50 + Aa 0.50)
	2	0.1250		0.0625 0.0625
Aa 0.50	1	(AA 0.25 + Aa 0.50 + aa 0.25)	+	(AA 0.25 + Aa 0.50 + aa 0.25)
	2	0.0625 0.1250 0.0625		0.0625 0.1250 0.0625
aa 0.25	1	(aa 1.00)	+	(aa 0.50 + Aa 0.50)
	2	0.1250		0.0625 0.0625

Frequency of heterozygous loci	0.375
Frequency of homozygous loci	0.626
	1.000

1 Frequency within progenies
2 Frequency in the total F_3 population

Fig. 2: Frequency of heterozygous loci in the F_3 progeny of a cross of 2 homozygous lines (AA and aa) when the rate of selfing is 50%.

insect pollination was prevented by an isolation cage, thus, in this part of the experiment, seed-set was only possible through spontaneous selfing, which may be defined as autofertility. In Experiment I the proportions of self- and cross-pollination were influenced by the amount of insect visitation to the flowers.

Yields from Experiment I are given in Table 1 and most interesting is the absence of heterotic effects in the F_2 progenies. This is clearly different from what would be expected from the results of the F_1 with quite well expressed levels of heterosis of 114, 122 and 126% of the parental mean. There were no significant differences between progenies and parental means. However, in this experiment, the errors were high as a result of aphid attack at the end of the growing period. Plants in Experiment II were grown in an isolation cage, and it is

TABLE 1

MEAN YIELD PER PLANT (g) AND AS PERCENTAGES OF THE PARENTAL MEANS

Experiment I (open pollinated)	Cross: L3 x L1		L7 x L1		L7 x L3	
	abs.	%	abs.	%	abs.	%
Generation						
P_1	36.4	111	36.4	110	29.6	101
P_2	29.6	89	29.3	90	29.3	99
$P_1 + P_2/2$	33.0	100	32.8	100	29.4	100
F_1	37.9	114	40.1	122	37.3	126
F_2	29.3	89	30.8	94	27.4	93
F_{3I}	33.1	100	26.5	81	25.5	87
F_{3II}	31.4	95	27.1	82	23.3	81
$\bar{x}$	32.9		31.7		28.7	
LSD 5%	ns (ns)[1)]		ns (ns)		ns (ns)	
1%	- (-)		- (-)		- (-)	
Experiment II (isolated, selfed)						
P_1	29.4	108	29.4	102	25.3	94
P_2	25.3	92	28.3	98	28.3	106
$P_1 + P_2/2$	27.3	100	28.8	100	26.8	100
F_1	37.2	136	37.9	132	37.8	141
F_2	33.6	123	30.8	107	32.3	121
F_{3I}	27.5	100	28.1	98	26.5	99
F_{3II}	28.8	105	28.1	98	26.0	97
$\bar{x}$	30.3		30.4		29.4	
LSD 5%	4.3 (5.0)[1)]		3.6 (4.2)		4.0 (4.6)	
1%	6.2 (7.1)		5.1 (5.9)		5.6 (6.5)	

1) Figures in ( ) refer to the parent with the highest value.

clear (Table 1) that the results are in accordance with the heterozygosity theory of heterosis i.e. a decrease in yield from F_1 - F_3 in relation to the increase of homozygous loci in the population.

Yields of F_1 progenies were significantly greater in all crosses than those of the best parent. Also, in two out of three cases, yields of the F_2 plants were markedly greater than those of the parents. Differences in numbers of seeds per plant and seed weight (Tables 2 and 3), in general, corresponded to differences in plant yield. In Experiment I there was no evidence of heterosis in numbers of seeds per plant in the F_2 although there was an indication of heterosis for this character in some F_1 crosses (i.e. L7 x L1 and L7 x L3). In contrast, there was a clear indication of heterosis for seed number in Experiment II, values in relation to parental levels decreasing from F_1 to F_3. Average seed weight in Experiment II (Table 3), however, tended to be lower for the F_1 and F_2 plants than for the parents. Even though compensation occurs between seed weight and seed number, the increased seed number was sufficient to give higher yields for the F_1 and F_2 plants.

It is stressed that these preliminary results should be treated with caution with regard to their interpretation, but the author knows of no similar investigations with which to make comparisons. However, support for these findings from elsewhere would suggest caution in the extensive use of synthetic populations, unless a clear understanding emerges of why the heterozygosity hypothesis is not valid for open-pollinated faba bean populations. In experiments reported here, genuine synthetics were not used, only two parents being used to build up a progeny. With two parents a heterotic character in F_2 would be expected to be reduced to 50% of the heterotic expression of F_1; with 3, 4 or 5 parents, which would conform more with a commercial synthetic, a decrease of heterosis of 1/3, 1/4 or 1/5 of the average F_1 expression could be expected for a set of diallel crosses. However, in all cases, the F_2 expression of the

TABLE 2

MEAN SEED NUMBER PER PLANT AND AS PERCENTAGES OF THE PARENTAL MEANS

Experiment I (open pollinated)	Cross: L3 x L1		L7 x L1		L7 x L3	
	abs.	%	abs.	%	abs.	%
Generation						
P_1	46.6	69	46.6	93	88.4	124
P_2	88.4	131	53.9	107	53.9	76
$P_1 + P_2/2$	67.5	100	50.3	100	71.2	100
F_1	67.6	100	53.4	106	82.9	116
F_2	60.7	90	48.3	96	64.7	91
F_{3I}	65.2	97	43.0	85	51.1	72
F_{3II}	63.6	97	44.1	88	55.6	78
$\bar{x}$	65.4	-	48.2	-	66.1	-
LSD 5%	14.4 (16.6)[1)]		ns (ns)		14.2 (16.4)	
LSD 1%	20.4 (23.6)		- (-)		20.3 (23.4)	
Experiment II (isolated, selfed)						
P_1	29.4	63	29.4	75	63.4	112
P_2	63.4	136	49.0	125	49.0	87
$P_1 + P_2/2$	46.4	100	39.2	100	56.2	100
F_1	57.8	124	47.3	120	77.9	138
F_2	54.2	116	40.9	104	67.5	120
F_{3I}	47.1	101	40.3	102	47.8	85
F_{3II}	46.3	99	36.7	93	43.6	77
$\bar{x}$	49.7	-	40.6	-	58.2	-
LSD 5%	8.2 (9.5)[1)]		5.5 (6.4)		8.5 (9.8)	
LSD 1%	11.7 (13.5)		7.9 (9.1)		12.0 (13.9)	

1) Figures in ( ) refer to the parent with the highest value.

TABLE 3

MEAN 1000-SEED WEIGHT (g) AND AS PERCENTAGES OF THE PARENTAL MEANS

Experiment I (open pollinated)	Cross:					
	L3 x L1		L7 x L1		L7 x L3	
	g	%	g	%	g	%
Generation						
P_1	789	145	789	122	299	75
P_2	299	55	494	78	494	124
$P_1 + P_2/2$	544	100	642	100	397	100
F_1	559	102	741	115	444	112
F_2	473	87	646	101	416	105
F_{3I}	508	93	597	93	528	133
F_{3II}	473	87	596	93	430	108
$\bar{x}$	516	-	643	-	435	-
LSD 5%	65 (75)[1]		82 (94)		68 (79)	
1%	92 (107)		116 (134)		97 (112)	
Experiment II (isolated, selfed)						
P_1	996	149	996	126	339	73
P_2	339	51	585	74	585	127
$P_1 + P_2/2$	667	100	790	100	462	100
F_1	636	95	786	99	482	104
F_2	632	95	740	94	458	99
F_{3I}	600	90	691	87	564	122
F_{3II}	627	94	727	92	607	131
$\bar{x}$	638	-	754	-	505	-
LSD 5%	34.5 (39.9)[1]		43.1 (49.7)		37.9 (43.8)	
1%	49.1 (56.7)		61.3 (70.8)		53.9 (62.3)	

[1] Figures in ( ) refer to the parent with the highest value.

heterotic character would be expected to be rather greater than that of the parental mean.

The results give some indication that heterosis in faba beans is mainly a result of differences in autofertility, which may be unimportant in the open field when insect pollination is adequate. Drayner (1959) showed that the genetic condition of the faba bean plant (heterozygous or homozygous) had a major influence on autofertility as hybrid plants are much more autofertile than inbreds. Interpreting these findings in relation to the heterozygosity theory of heterosis, it would be expected that maximum autofertility should occur in F_1 and the minimum in the inbred parents; F_2 and F_3 plants would be intermediate between F_1 and the parental mean. Autofertility, defined as the ratio of number of pods to number of flowers (Table 4), shows no consistent trends in the various crosses. Lines L3 and L1 were of low autofertility and their F_1 and F_2 generation crosses showed a higher degree of autofertility than the parental mean or the best parent. In both crosses involving L7, with its high level of autofertility, the F_1 and F_2 of the L7 x L1 cross showed lower autofertility than that of the parent with the highest autofertility. In the L7 x L3 cross autofertility of the F_1 and F_2 progenies was similar to that of line 7.

A summary of the averages for yield, number of beans per plant and average seed weight are given in Table 5 and show only minor differences in plant yield between the experiments. This would suggest that opportunity for fertilisation was similar for plants in each experiment, and it is concluded that the responses obtained were due to differences in pollination rather than differences in environment.

As mentioned earlier, the seed numbers in Experiment I were lower than those in Experiment II and were probably a result of suboptimal pollination or fertilisation. Nevertheless, these lower numbers of seeds were compensated for by increased average seed weights.

TABLE 4

AUTOFERTILITY = RATIO: $\frac{\text{NO. OF PODS}}{\text{NO. OF FLOWERS}}$ OF 3 NODES

Generation	Cross: L3 x L1	L7 x L3	L7 x L1
P_1	34	34	38
P_2	38	73	73
$P_1 + P_2/2$	36.0	53.5	55.5
F_1	49	60	74
F_2	50	42	79
F_{3I}	39	50	68
F_{3II}	40	45	52
$\bar{x}$	41.7	50.6	64.0

TABLE 5

TOTAL MEANS OF EXPERIMENT I AND II

	Cross:	L3 x L1	L7 x L1	L7 x L3
No. of seeds/plant	Exp. I:	65.4	48.2	66.1
	Exp. II:	49.7	40.6	58.2
1000-seed weight	Exp. I:	516	643	435
	Exp. II:	638	754	505
Yield/plant	Exp. I:	32.9	31.7	28.7
	Exp. II:	30.3	30.4	29.4

DISCUSSION

D.A. Bond *(UK)*

Was the parent L_7 used by Dr. von Kittlitz the auto-fertile one, was it highly bred and what was it's origin?

E. von Kittlitz *(FRG)*

L_3 was a minor type and L_7 was an equina type and L_1 was a major type. They were very different in some characters such as branching, pod shape and plant height. The genetic background of the lines differed markedly.

D.A. Bond

You ask if others have had similar results. If you are asking whether other people found auto-fertile inbred lines then, yes, Dr. Poulsen has one and we have one; auto-fertility is not just a question of heterozygosity.

E. von Kittlitz

I wanted to know if you have any results that show F_2 generations have no indication of heterosis. I thought it surprising that I could not find any indication of heterosis in F_2 and F_3 generations, in spite of the fact that we found heterosis in F_1.

D.A. Bond

I cannot answer that very well, but from the data I agree it is surprising. The F_2 data was intermediate between the F_1 and the inbreds; there was some residual heterozygosity in F_2.

D.A. Lawes *(UK)*

I am still not sure what Dr. von Kittlitz is asking. Are

you asking about the heterozygosity effect on auto-fertility or on open-pollinated yield in the field?

E. von Kittlitz

I think it is important to have heterosis in the open field. The astonishing thing is that we can demonstrate the degree of heterosis to be expected from the heterozygosity present but not in the open field.

D.A. Lawes

Is it not possible, under field conditions, that if you have a high level of bumble bee activity you do not require auto-fertility in the field? Is it really so surprising?

E. von Kittlitz

No, it is not surprising, but from the viewpoint of breeding principles there was no information on levels of heterosis.

D.A. Lawes

As far as I am concerned we look upon auto-fertility as a mechanism to make yields more reliable in the absence of bumble bees. If bumble bees are present then the yields are satisfactory anyway. The value of auto-fertility is where self-pollination or self-fertilisation is inadequate. Personally, I see no problem; we are not looking for higher potential yield, we are looking for a more reliable yield.

G. Dantuma *(Netherlands)*

One of the problems is that we are trying to find a general rule, and there is no general rule. There are enormous differences in inbreeding effects between varieties.

R. Thompson *(UK)*

That is a very valid point and one on which we should perhaps terminate this part of the discussion for the time being. The point made by Dr. Lawes is also valid: auto-fertility is a safety valve.

G.C. Hawtin *(Syria)*

I would like to ask one question on the index of auto-fertility defined as the number of pods divided by the number of flowers. One can only use that as an index of auto-fertility providing you have parent lines which have the same number of flowers. Can the type which produces large numbers of flowers but a small pod set be said to be auto-fertile?

E. von Kittlitz

We restricted our observations on flowers to 3 flowers per node.

J.J. Cubero *(Spain)*

With 4 - 5 generations you have the two parents of F_1, F_2 and F_3. With mathematical methods it should be possible to see whether there is heterosis or not. It can be computed very easily but you need the variances of any generation; for simplicity, take the variances of the parents.

I believe that three nodes is too few. When you had many nodes did you measure the repeatability of the auto-fertility index character? The repeatability of the flowers per node is high, but the repeatability of the pods per node is very low. In your material, what was the repeatability of the auto-fertility index - pods divided by flowers? It is necessary for this to be a stable character; a repeatable character is not the same. The number of flowers is important when varieties differ for this character. There is a way to resolve this: fertility can be related to the proportion of young pods (of fertilised flowers) divided by the number of flowers. It is more time-consuming but more accurate because you can be sure that all the flowers have fully developed. Then there is pod setting. If you have a line, a hybrid or variety with, for example, seven flowers, these may not all develop into pods even though the variety is self-fertile. Pod drop may result from competition within the plant.

R. Thompson

I understand you also to mean that the great difficulty is in determining which is the young pod?

J.J. Cubero

Immediately after ferilisation of the flower the petals drop, and a pod producing seeds can then be identified. At a particular stage in faba beans the young pods may drop for physiological reasons. Up to 8 or 9 flowers may produce pods initially but not all are retained to become mature pods. I have found that the correlation coefficient between these physical characteristics is very high. The important point is that not all of the fertilised flowers produce mature pods. The index for self-fertility should preferably be young pods divided by the number of flowers.

M.H. Poulsen *(Denmark)*

I agree with Professor Cubero in general. But, in his method, fertilisation of the flower must be ensured, because in some lines flowers develop pods without fertilisation.

R. Thompson

Thank you Dr. Poulsen and Dr. von Kittlitz, we must now move on but flower abscission will be dealt with more fully this afternoon.

SESSION 3

FLOWERING, POLLINATION AND POD/SEED SET

Chairman: J. Picard

STUDY OF THE FERTILITY COMPONENTS IN FABA BEANS (*Vicia faba* L.) - VARIABILITY AMONG SIX DIFFERENT GENOTYPES - EFFECT OF TOP AND FLOWER REMOVAL

G. Duc and J. Picard
Station d'Amelioration des Plantes, INRA,
21034 Dijon Cedex, France

ABSTRACT

Stem apex and flower removal treatments were applied to six morphologically very different genotypes. Two independent sets of characters were observed on the plants: a flower abcission character (YP/F) and a pod nutrition character (P/YP, S/P, W).

Top removal reduced flower abcission but resulted in poor pod nutrition. Flower removal also reduced flower abcission but improved pod nutrition. Yield was decreased and protein content increased by the treatments in comparison with the control. Genotype x treatment interaction concerned more the magnitude of response than the type of response to treatment. The results identified within-plant competition between organs including:

- *between the top of the plant and young pod formation.*
- *between flowers on a node destined to become young pods,*
- *between pods in their relation to the leaf area, as a source of nutrients.*

The significance of this within plant competition is discussed.

INTRODUCTION

As a result of low and unstable yields, the area of cultivated faba beans in Europe has not increased during recent years although importation of soya bean meal has been increasing.

The breeding programmes aimed at improving the total yield of this crop have not made very significant progress so far. Therefore an examination was started of the yield components, (numbers of flowers, young pods (1 cm), pods and seeds per plant). The purpose was to obtain a better understanding of the factors limiting development of the following characters:

Young pods/flower (YP/F)
Pods/young pod, (P/YP)
and seeds/pod. (S/P)

We used two types of approach for this study,

- a physiological approach, modifying plant form by removing the apex and flowers,
- a genetic approach, examining the response of six morphologically very different (Figure 1) genotypes to these treatments.

The following six genotypes were used in each of two field trials with the plant density adjusted to 25 plants/m^2.

A Line 319 (Dijon)
B Line 370 (Dijon)
C Line 247 (Dijon)
D Hybrid (Dijon)
E Ascott
F Blaze

Each trial comprised four replications.

The first trial was used to measure the harvest index (ratio of the dry seed yield to the maximum dry matter production of the aerial part of the plant during vegetative growth).

Cultivar A | Cultivar B

number of tillers 2.08 | 1.26

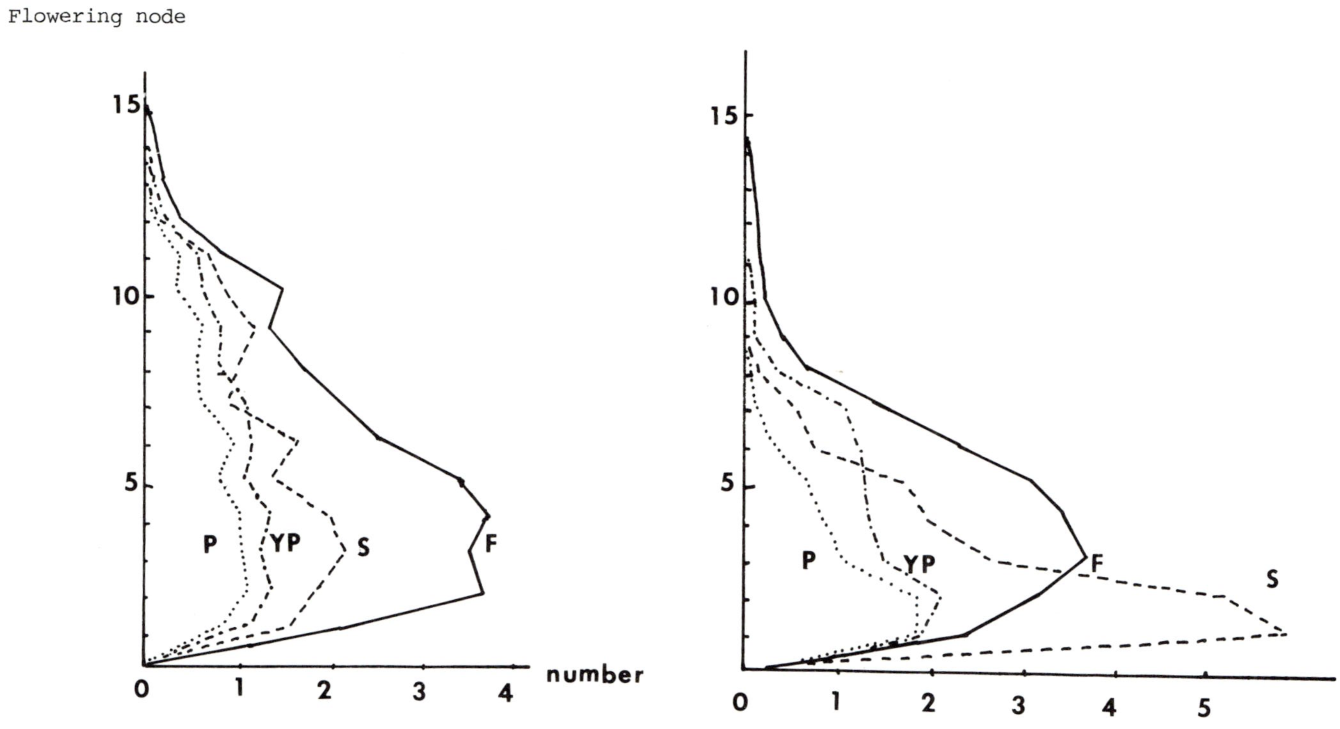

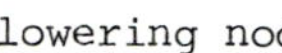

Fig. 1: Main stem: profile of the 6 cultivars.
Number of flowers/node = F; young pods/node = YP; pods/node = P; seeds/node = S. (Continued).

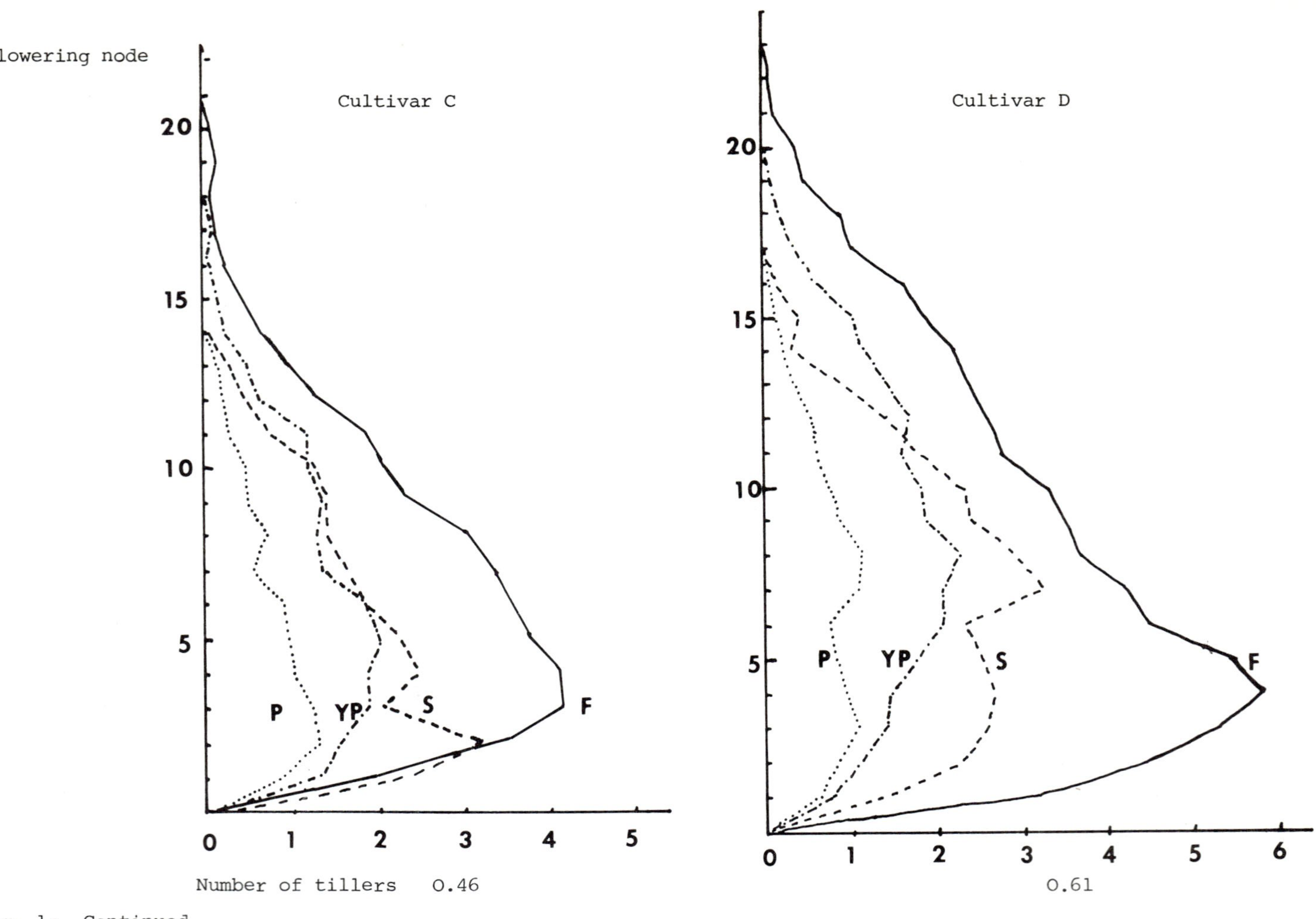

Fig. 1: Continued.

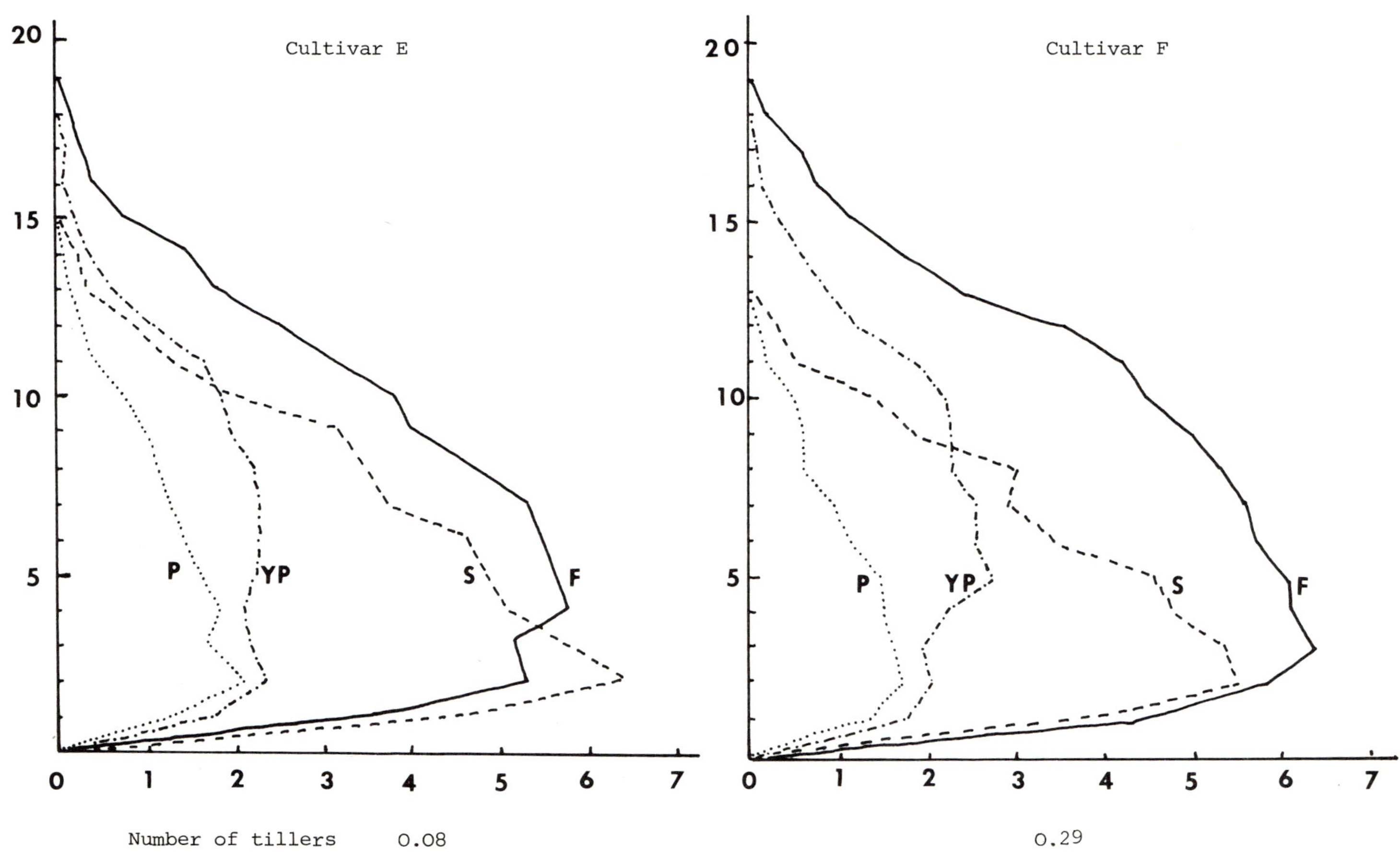

Fig. 1: Continued.

In the second trial the effects of the following treatments were assessed:

Top removal after anthesis of the first 5 flowering nodes

T 1 - all flowers removed on tillers,

T 2 - all tillers removed,

Top removal after anthesis of the first 10 flowering nodes

T 3 - all flowers removed on tillers,

T 4 - all tillers removed,

T 5 - all flowers removed after anthesis of the first 5 flowering nodes,

T 6 - all flowers removed after anthesis of the first 10 flowering nodes,

T 7 - 10 flowers left on the first 10 flowering nodes. All others removed,

T 8 - 20 flowers left on the first 10 flowering nodes. All others removed,

T 9 - 40 flowers left on the first 10 flowering nodes. All others removed,

T 10 - Control.

Notation of the flowering nodes was according to their chronological order of anthesis. Treatments were applied to 6 plants in each plot. Some irregularities in the final density of plants did not allow the replication of treatments T 1, T 2, T 3 and T 4 to the genotypes A and D, and T 5 and T 6 to the genotype D. Allowance was made for this by modifying the partition of the effects in the analysis of the results.

Main characters of the 6 genotypes

The six genotypes were lines, hybrids and cultivars very different in origin and morphology.

Their difference in flowering date (Table 1) probably had little effect on the yield components as the climatic conditions were quite stable during the flowering period.

TABLE 1

MAIN CHARACTERS OF THE 6 GENOTYPES

	A	B	C	D	E	F	Max/Min
Flowering date	22/5	26/5	23/5	27/5	31/5	31/5	
Level of the first flowering node on the main stem	4.23	5.7	5.09	5.82	7.16	6.49	1.69
Number of tillers/plant	2.08	1.26	0.46	0.61	0.08	0.29	26
Leaf area/plant (cm^2)	1878	2257	1695	3085	2138	3078	1.64
Number of flowering nodes/plant	23.5	16.6	17.7	22.1	14.1	16.7	1.67
Number of flowers/plant	55.4	39.6	47.0	73.7	60.7	80.9	2.04
Number of young pods/plant	26.4	20.6	24.2	32.1	25.5	32.0	1.56
Number of pods/plant	17.8	12.7	11.0	12.9	15.1	13.2	1.62
Number of podded nodes/plant	14.9	8.4	9.1	9.7	9.1	9.3	1.77
Number of seeds/pod	1.96	2.54	2.39	2.65	3.09	3.13	1.60
Mean weight of a seed (g)	0.52	0.80	0.65	0.55	0.39	0.41	2.05
Yield per plant (g)	18.1	25.9	17.1	18.7	18.3	16.8	1.54
Harvest index	44.6	48.5	45.7	37.8	44.3	40.9	1.28
Protein content of the seeds (% DM)	32.7	27.4	30.4	32.5	32.0	31.2	1.19
YP/F* (%)	47.7	52.1	51.5	43.6	42.0	39.5	1.32
P/YP (%)	67.3	61.6	45.5	40.0	59.1	41.2	1.68
Flowers/flowering node	2.36	2.39	2.66	3.34	4.31	4.86	2.06
Seed yield (g) per podded node	1.21	3.08	1.88	1.93	2.01	1.81	2.55
Number of lower nodes on the main stem carrying 75% of the yielld of the stem	5	3	5	8	7	6	
Proportion of yield on the main stem	44.7	56.8	82.0	83.3	98.5	90.6	

* YP/F = Young pods/flower
* P/YP = Pods/YP

Marked variability occurred for several characters (Table 1) such as: number of tillers/plant, number of flowers/plant, number of flowers/flowering node, weight of a seed and seed yield/podded node.

A very low variation was observed for the harvest index

genotype D, which was a hybrid incompletely restored.

Examination of the distribution of the yield on the plant, shows that the proportion of the yield on the main stem is very variable. The profiles of the main stem for F, YP, P and S (Figure 1) show variability for the number of flowering nodes, the number of seeds per node and the level of concentration of the yield on the lower nodes.

For the 6 genotypes, the characters YP/F and P/YP had low values with averages of 46% and 52% respectively, and the variability for P/YP was higher than for YP/F.

RESULTS

Two groups of characters can be distinguished with regard to their correlations. The first is designated 'Pod nutrition characters' and includes P/YP, S/P and seed weight (W). A high positive correlation was measured between each of these characters in response to the treatments.

The second is designated 'Flower abscission character' which is YP/F and appears to be independent of the first.

In the control plants, YP/F had a higher value on the upper part of the main stem and on the tillers than on the lower part of the main stem. The reverse was true for P/YP and S/P. Correlations between flower abscission and pod nutrition characters were positive for some treatments but negative for others depending on the particular treatment imposed.

These two groups of characters were analysed separately.

MAIN EFFECTS OF THE TREATMENTS

Effects on flower abscission (Table 2)

Top removal considerably reduced flower abscission. The proportion of flowers setting young pods was increased by about

TABLE 2

EFFECT OF THE TREATMENTS ON THE FERTILITY COMPONENTS

Fertility component / Treatments	YP/F	% of the control	P/YP	% of the control	S/P	% of the control	Weight of a seed (g)	% of the control
1	0.659] NS	139.6	0.356x NS	58.0	2.543	90.8	0.593x NS	105.5
2	0.700 NS	148.3	0.289	47.1	2.301	82.1	0.442	78.65
3	0.581]	123.1	0.368 x] NS	59.9	2.484] NS	88.7	0.513] NS	91.28
4	0.597]	126.5	0.354] x	57.7	2.438]	87.0	0.503]	89.5
Control	0.472	100.0	0.614	100.0	2.801	100.0	0.562x	100.0
5	0.479]	103.9	0.806	125.3	2.869	109.0	0.695	125.45
6	0.479] NS	103.9	0.670] NS	104.2	2.683] NS	101.9	0.565] NS	101.99
Control	0.461]	100.0	0.643]	100.0	2.633]	100.0	0.554]	100.0
	(1) (2)	(1) (2)						
7	0.233 0.855	52.1 191.3	0.847	135.5	2.741]	103.7	0.700	126.58
8	0.379 0.733	84.8 164.0	0.721	115.4	2.798] NS	105.9	0.604] NS	109.22
9	0.457] NS 0.495	102.2 110.7	0.656] NS	105.0	2.688]	101.7	0.585] NS	105.79
Control	0.447]	100.0 100.0	0.625]	100.0	2.642]	100.0	0.553]	100.0

(1) Initial number of flowers before treatment
(2) Final number of flowers after treatment

F = Flowers: YP = Young pods: P = Pods: S = Seeds.

40% in T 1 and T 2 and 25% in T 3 and T 4. Removing the tillers had no effect.

Flower removal also improved setting of young pods with 91% and 64% increases in T 7 and T 8 respectively. The other treatments, T 5, T 6 and T 9 gave similar values for pod set to that of the control and did not have any significant effect.

It was noticeable that T 7 led to an average young-pod set of 85.5%, reaching 94% for genotype F. This indicated a high general level of pollination in the trial.

Effects on pod nutrition (Table 2)

Top removal reduced P/YP by 40 - 50%, S/P by 10 - 20% and W by 10 - 20%. Removal of tillers from the plants from which tops were removed at 5 nodes (T 2) reduced the values for the pod nutrition factors.

In general flower removal resulted in increased values for the pod characters, excluding those for T 6 and T 9. For these two treatments, values for flowers per node and number of podding nodes did not differ markedly from those of the control.

Treatments T 5, T 7 and T 8 increased P/YP by 15 - 35%, S/P by 4 - 9% and W by 9 - 27%. These three treatments induced a concentration of the yield on fewer nodes. Plants from T 5, for example, had a seed yield per podded node of 158% of that of the control. Similar numbers of pods were formed in T 5 and T 10 even though only half the number of nodes were allowed to set pods in the former compared with the latter. Nevertheless the value for P/YP was somewhat greater for plants from P 10.

Effects on the yield components (Table 3)

The characters that were considered were the number of pods/plant, seeds/plant, the seed yield per plant. All treatments reduced the values for these yield components.

Seed yield per plant was reduced by up to 80% by top removal

TABLE 3

EFFECTS OF THE TREATMENTS ON THE YIELD COMPONENTS

Yield components / Treatments	Number of pods/plant		Number of Seeds/plant		Weight of a seed (g)		Seed yield/plant		Protein content of the seeds % of the DM	
		% of the control		% of the control		% of the control		% of the control		% of the control
1	4.775 NS	37.67	12.281 x	34.21	0.593 x	105.5	6.420	33.59	31.268	103.43
2	4.111	32.43	9.630	26.83	0.442	78.65	3.805	19.91	31.450 NS	104.03
3	8.209 NS	64.76	20.822 NS	58.01	0.513 NS	91.28	9.450 NS	49.44	31.256	103.39
4	7.593	59.90	18.886 x	52.61	0.503	89.50	8.468	44.30	30.943 NS	102.36
Control	12.676	100.0	35.896	100.0	0.562 x NS	100.0	19.114	100.0	30.231	100.0
5	7.857	47.39	21.258	59.60	0.695	125.45	13.977 NS	73.99	32.130	104.57
6	11.025	80.53	29.234	81.96	0.565 NS	101.99	15.536	82.24	30.980 NS	100.83
Control	13.691	100.0	35.667	100.0	0.554	100.0	18.891	100.0	30.725	100.0
7	7.215	53.29	18.667	52.97	0.700	126.58	12.695	67.56	33.008	106.41
8	9.967	73.61	27.059 NS	76.78	0.604 NS	109.22	16.045 NS	85.38	31.129	100.35
9	11.271	82.24	29.527	83.78	0.585 NS	105.79	16.477	87.68	31.087 NS	100.22
Control	13.540	100.0	35.243	100.0	0.553	100.0	18.792	100.0	31.020	100.0

and up to 32% by flower removal. The number of pods per plant was reduced by up to 68% by top removal and up to 47% by flower removal. The number of seeds per plant was reduced by up to 73% by top removal and up to 47% by flower removal.

Effects on protein content (Table 3)

In general the effect of all treatments was to increase protein content, which was higher with flower removal than with top removal. In the case of top removal, the increase was associated with a decrease in the seed weight and a marked reduction in yield. In the case of flower removal, the increase in protein content was associated with an increased seed weight and a small reduction in the yield.

Genotype x treatment interaction

There were very few differences between genotypes in the type of response to the treatments but some differences in magnitude were detected (Table 4).

Interaction for flower abscission

Genotypes A and B showed only a slight response to the treatments for this character. These genotypes were also characterised by a high value of YP/F and a low number of flowers per flowering node on the control plants.

Interaction for pod nutrition

The character P/YP was slightly influenced by the treatments for genotypes A and B; of all six genotypes, values of P/YP for the control plants were greatest for genotypes A and B. Genotype F showed the greatest range in response to the treatments with values of from 34 to 169% of the control. Genotype E was very stable for the character S/P in response to treatment (±5% of the control) and cultivar C showed the greatest range of response (75 to 135% of the control).

TABLE 4

RESPONSE OF THE SIX GENOTYPES TO TWO OF THE MORE EXTREME TREATMENTS: T 2 FOR TOP REMOVAL, T 7 FOR FLOWER REMOVAL, EXPRESSED IN PERCENT OF THE CONTROL OF EACH GENOTYPE

Genotypes	Treatments	YP/ flower	Pod/ YP	Seed/ pod	Pod/ plant	Seed/ plant	Weight of a seed	Yield/ plant	Protein content (% DM)
A	T 2	-	-	-	-	-	-	-	-
	T 7	72.2	102.2	104.5	43.7	45.9	136.2	62.9	106.9
B	T 2	112.6	92.7	93.9	33.9	24.3	75.4	18.3	95.3
	T 7	53.7	57.3	73.8	57.6	57.1	128.5	72.9	110.6
C	T 2	158.5	136.9	134.1	28.6	20.7	87.8	17.8	102.5
	T 7	56.5	40.4	73.9	59.0	65.0	121.6	79.1	104.1
D	T 2	-	-	-	-	-	-	-	-
	T 7	49.2	123.9	86.6	45.0	39.1	115.9	44.4	107.5
E	T 2	169.5	126.5	104.0	37.9	35.8	77.0	27.2	102.7
	T 7	44.1	48.5	94.9	57.3	53.7	135.3	71.5	103.7
F	T 2	161.8	169.3	109.9	28.1	22.7	72.6	16.5	114.0
	T 7	38.1	34.0	82.6	61.5	58.5	125.3	74.0	105.6

Interaction for yield components

On the basis of the response to the treatments, genotypes B and C were very similar. Among the flower removing treatments, the yield per plant for genotype D was very much decreased by T 7.

Interaction for protein content

Cultivars B and F showed the greatest response to the treatments. But they showed contrasting responses to top removal, the protein content of seed from T 2 decreasing for genotype B to 95% of the control and increasing for genotype F to 114% of the control.

DISCUSSION

Each of the six genotypes included in the study showed different but high levels of flower and young pod abscission. The problem of abortion in *Vicia faba* has already been discussed in some detail (Kambal, 1969), and is known to occur widely in other legumes such as soyabean, *Phaseolus*, peas, lupins and cow-peas.

As already observed by Jacquiery (1977) pollination is not the sole determinant of flower drop. In our trial, high values YP/F of up to 94% were found with flower removing treatments.

The abortion of reproductive organs appeared to be a physiological response and the treatments applied caused a wide range in the levels of competition between the various organs.

Evidence of competition between the vegetative growth of the top of the plant and young pod formation was demonstrated by the top removal treatment. In similar experiments, Hodgson and Blackman (1957), Gehriger, (1978) and Chapman et al. (1978), have also shown an increase in the character YP/F. Although Jacquiery (1977) showed that the plant apex competed with the young embryos for assimilate, the involvement of a growth substance produced by the apex cannot be ignored in explaining flower abscission.

Competition exists between the flowers on the same node for assimilate for young pod growth. This was demonstrated by the flower removal treatment where the earliest flowers on a node were the more likely to develop into young pods. Kambal (1969) also reported an intra- and inter- ovary competition which favoured the development of the earliest embryos. This competition could be the result of growth inhibitors or promoters synthesised by the embryo.

The ratio of the leaf area to the number of young pods may be influential in determining abortion of the young pods. Intra-node competition exists in the development of young pods into pods. The artificial reduction of young pods on a plant by flower removal is compensated for by a higher level of nutrition for the remaining young pods.

Top removal treatments which enhanced the ratio YP/F also decreased the leaf area. The consequence was a higher level of abortion of young pods and lower yield. This phenomena was also reported by Hodgson and Blackman (1959), Gehriger (1978) and Chapman et al. (1978).

Translocation of assimilates seems to play a major role in pod nutrition as indicated by the effects of the apex and of the tillers on seed yield at the lower nodes. Moreover, the yield per node can be increased if the number of podded nodes is reduced to five, for example, probably because of a higher concentration of assimilates at each reproductive node.

These results suggest new approaches, some of which have already started at the INRA Plant Breeding Station of Dijon. For example:

Breeding to improve the ratio YP/F: the variability for this character in *Vicia faba* seems large and studies of its inheritance will be made.

Decreasing the intra-plant competition in young pod nutrition: our results suggest that types of interest include those with a short flowering period, few flowering nodes distributed on several stems and with few flowers per node.

Studies will be made on i) the interaction of genetic factors and physiological treatments in a wide range of phenotypes in respect of the components of yield. ii) Leaf area as a source of assimilate, including size, duration and shape. iii) The level of translocation of the assimilates from the leaves to the seeds, and, iv) Harvest index.

REFERENCES

Chapman, G.P. et al. 1978. Top removal on Single stem plants of *Vicia faba* L. X. Pflanzenphysiol. Bd. 89 S. 119-127

Gehriger, W. 1978. Influence de la temperature et de 'écimage sur le developpement de la feverole *(Vicia faba* L) et étude de la nutrition des fleurs en assimilats marqués au ^{14}C. Thesis No. 6133 - Ecole Polytechnique Federale - Zurich

Hodgson, G.L. and Blackman, G.E. 1957. An analysis of the influence of plant density on the growth of *Vicia faba*. II. The significance of competition for light in relation to plant development at different densities J. Exp. Bot 8, 195-219

Jacquiery, R. 1977. Etude de la chute chez la feverole *(Vicia faba* L*)* relations avec la disonibilité en assimilats marqués au ^{14}C. Thesis No. 5893, Ecole Technique Federale, Zurich

Kambal, A.E. 1969. Flower drop and fruit set in field beans. J. Agric. Sci. Camb. 72, 131-138

ACKNOWLEDGEMENT

We wish to thank Francois Girard, (student) and Mms. Berthaut (technician at the Laboratory) for their important contribution to this work.

CELLULAR CHANGES IN THE PEDICEL AND PEDUNCLE DURING FLOWER ABSCISSION IN *Vicia faba*

P. Gates, Jennifer N. Yarwood, N. Harris, M.L. Smith and Eileen Boulter

Botany Department, University of Durham, Durham, UK.

ABSTRACT

Changes in the cellular structure of the pedicel and peduncle of Vicia faba *L. during early pod setting and flower shedding have been followed using the light and electron microscopes.*

Early in pod development, a rapid proliferation of vascular tissue and thickening of vascular pith parenchyma cells occurs, giving the junction of the peduncle and pedicel great mechanical strength.

During flower shedding, lysis of the middle lamellae occurs across a layer of cortical cells at the peduncle/pedicel junction, followed by expansion of the cells on the proximal side of the abscission zone, fracturing the vascular connection.

A hypothesis is presented to account for the observed pattern of flower shedding in axillary racemes.

INTRODUCTION

Shedding of reproductive structures in *Vicia faba* L., which can reach levels of 97% of flowers initiated, has been described by many authors (for review, see Kambal, 1969); it is a common phenomenon in many wild and cultivated legumes. The number of fertilised flowers retained by the plant (flower set) determines the number of pods potentially capable of reaching maturity and hence overall crop yield and stability.

A distinction must be drawn between flowers which are shed, fertilised or unfertilised, before they begin to develop into pods, and those which are shed during early pod formation. In the latter case competition for assimilates between sinks, at the apex and in the seeds, has been shown to be the most important factor in the failure of pods to reach maturity (Jaquiéry and Keller, 1978a, b).

This paper, on the other hand, compares the cellular development of the pedicel and peduncle of *V. faba* during the earliest stages of pod development, and during flower shedding under stress-free environmental conditions before a young pod is formed and becomes an active sink. Studies in cowpea have shown that until this stage is reached, availability of nutrients is of secondary importance in determining flower abscission (Ojehomon, 1972).

MATERIAL AND METHODS

Plants of *V. faba* inbred lines STW (autofertile, few-flowered racemes, Sudanese Triple white) and T2 (autosterile, Northern European, with many-flowered racemes) were grown in 5" pots of John Innes compost in glasshouses during April and May 1980. All flowers were tripped at anthesis by depressing the keel petal. Previous observations had shown that under these conditions, basal flowers on the racemes of STW always developed into pods, whereas terminal flowers on the racemes of T2 were almost invariably shed. Flowers removed from these positions could be classified as either setting or shedding with a high degree of confidence.

For thick sectioning, flowers were fixed in formalin : ethanol acetic acid (1 : 8 : 1) for 1 h, dehydrated in an alcohol series, transferred to xylene, embedded in wax and cut on a rotary microtome at 12.5 μm. Sections were rehydrated through an alcohol series and stained in either Toluidene blue, periodic acid-Schiff's reagent, I/KI or acetocarmine.

For thin sectioning, material was fixed for 2 h in 2.5% glutaraldehyde and 1.0% formaldehyde in pH 7.0 0.1 M cacodylate buffer, washed in buffer (2 h), postfixed in 1% aqueous osmium tetroxide (3 h), washed in water, dehydrated through an alcohol series, embedded in Spurr's resin, and sectioned at 1 μm in an ultramicrotone.

Thin sections for transmission electron microscopy were stained in uranyl acetate and lead citrate and examined in a Phillips EM2000 microscope.

Specimens for scanning electron microscopy were prepared by fixing in glutareldehyde and dehydrating in alcohol, after which they were transferred to acetone, critical point dried with CO_2, coated, and examined in a Cambridge S 600 SEM.

Ovaries were examined for the presence of pollen tubes by the method of Martin (1959).

OBSERVATIONS

(A) Morphology of pedicel and peduncle at anthesis

The vascular supply consisted of four vascular bundles, three large and one small, surrounded in the pedicel by a narrow ring of approximately six ranks of cortical cells. At anthesis, large reserves of starch were present in a sheath around the vascular bundles and in the parenchyma cells between them (Figure 1).

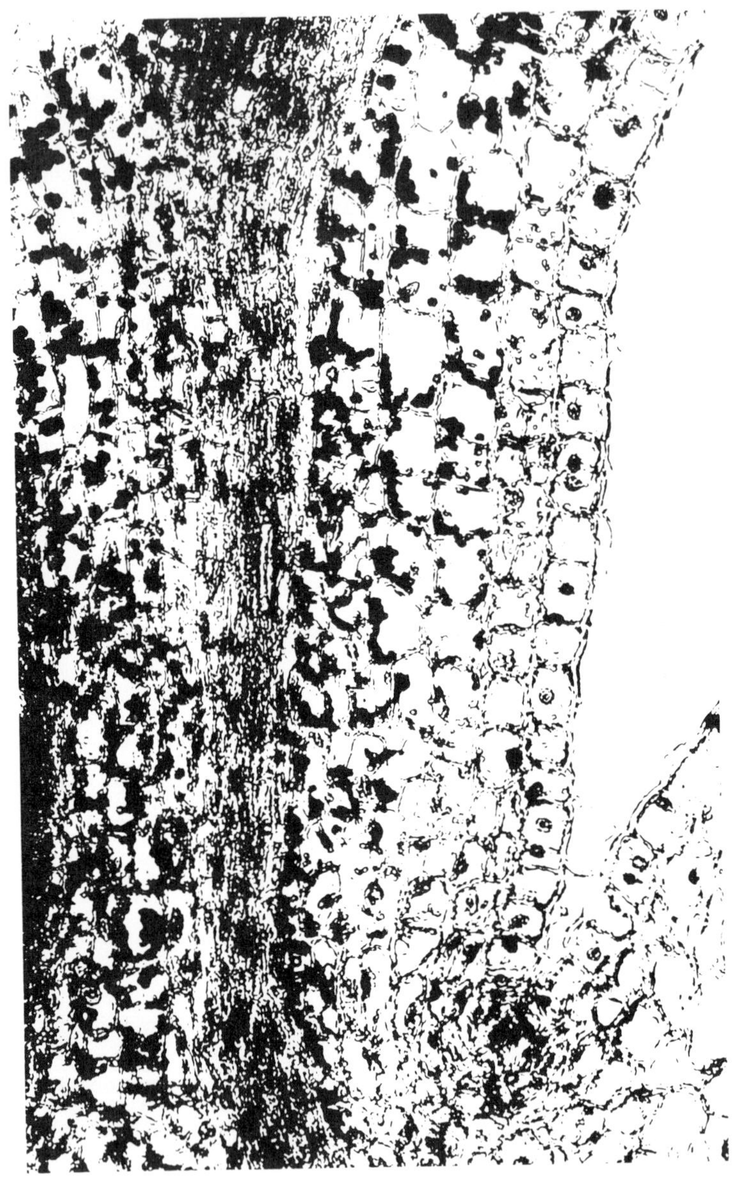

Fig. 1. Longitudinal section through base of pedicel, stained with I/KI to show starch accumulation around the vascular bundles. x 60.

At the junction of the peduncle and pedicel a ring of small, thick-walled meristematic cells with dense cytoplasm surrounded the vascular bundles.

The cellular structure of the pedicel and peduncle is shown diagrammatically in Figure 2.

B. Changes during early pod set

Between anthesis and pod maturity in line STW, the pedicel rapidly increased in diameter, from 1 mm just prior to pollination to 6 mm at the time of maximum pod fresh weight. In the first three days after fertilisation cortical cells of the pedicel increased in mean diameter from 45 μm to 70 μm, concurrently with the swelling of the ovary.

Cell division and expansion occurred in the ring of meristematic cells at the peduncle/pedicel junction, and the vascular bundles rapidly increased in size so that at the base of the pedicel they formed an almost complete cylinder of thickened vascular tissue, penetrated by narrow medullary rays of parenchyma cells. The cross-sectional area of vascular tissue increased four-fold in the period between anthesis and the formation of a pod 2 cm long.

At the 2 cm pod stage in line STW the parenchyma cells within the vascular cylinder showed a characteristic wall thickening (Figure 3), and by this stage the starch present at anthesis had disappeared.

C. Changes during flower shedding

Sectioning of the peduncle/pedicel junction at intervals throughout flower development showed that the first obvious morphological change preceding flower shedding was the appearance of an annular split at the base of the pedicel. This occurred shortly before flower shedding and extended from the epidermis to the vascular tissue, so that the flower remained attached to the peduncle solely by the vascular bundles

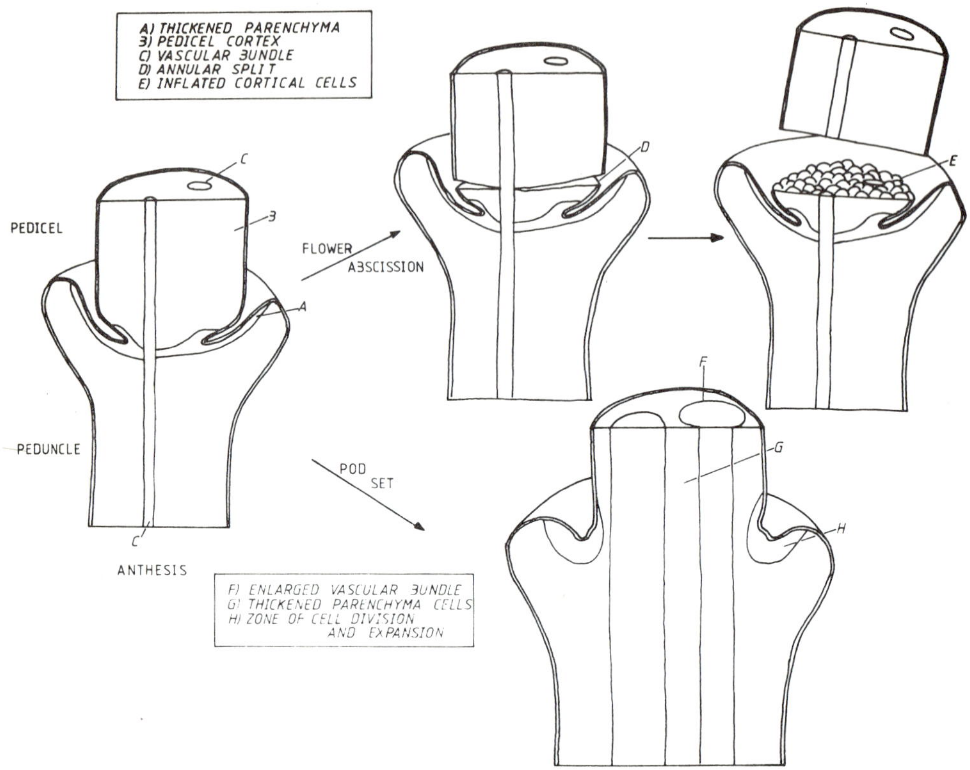

A) THICKENED PARENCHYMA
B) PEDICEL CORTEX
C) VASCULAR BUNDLE
D) ANNULAR SPLIT
E) INFLATED CORTICAL CELLS

PEDICEL

PEDUNCLE

ANTHESIS

FLOWER ABSCISSION

POD SET

F) ENLARGED VASCULAR BUNDLE
G) THICKENED PARENCHYMA CELLS
H) ZONE OF CELL DIVISION AND EXPANSION

Fig. 2. Diagrammatic longitudinal sections of the peduncle/pedicel junction, showing arrangement of tissues at anthesis and alternative developmental pathways.

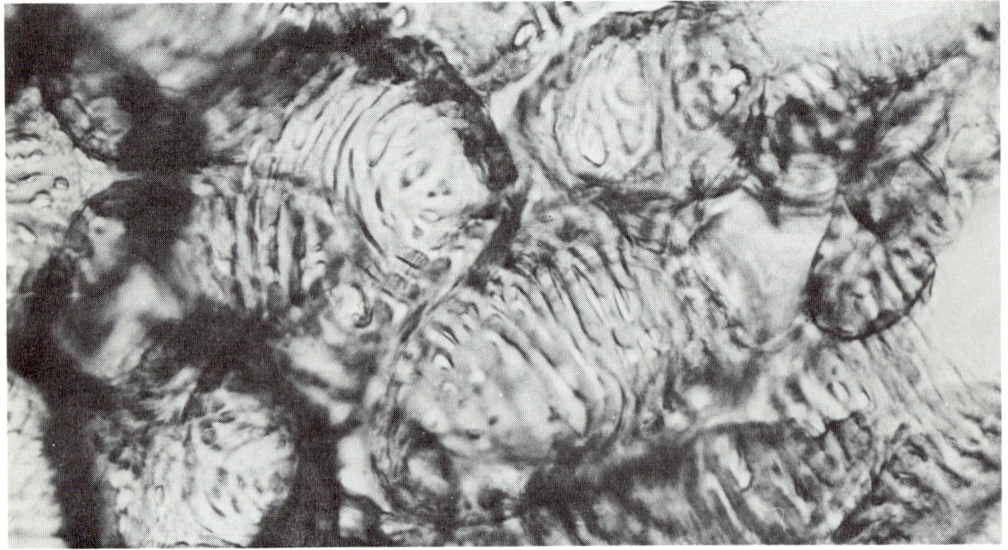

Fig. 3. Wall thickening in the vascular pith parenchyma at the 2 cm pod stage. x 150.

(Figure 4a, b). Ultrastructural examination showed that the upper ranks of thickened parenchyma cells had separated along their middle lamellae (Figure 5).

SEM examination of the abscission region immediately after flower shedding revealed a ring of inflated cortical cells which had pushed the flower off after rupturing the vascular connection (Figure 6). Light microscope sections and the SEM showed spiral thickening of the xylem often visibly protruding from the abscission zones. The coils of the xylem thickening were expanded, showing that the swelling of the cortical cells had stretched the vascular connection until it had finally ruptured.

Sexton (1976) observed similar cellular changes in *Impatiens sultani* leaf abcission zones. Following flower abcission, the inflated cortical parenchyma cells collapsed, and continued meristematic activity led to the production of new epidermis to cover the scar on the peduncle left by the abscising flower.

D. Fertilisation in abscised flowers

The proportion of flowers with pollen tubes detected in the ovary is given in Table 1, together with the number of flowers sampled for each genotype.

TABLE 1

FREQUENCY OF ABSCISED FLOWERS WITH POLLEN TUBES DETECTED IN THE OVARY

Variety/line	No. flowers sampled	% flowers with pollen tubes
In the field:		
Compacta (commercial variety)	70	22.9
NS 74 (autofertile inbred line)	159	39.6
T51 (autofertile inbred line)	56	16.1
51/3 (autosterile inbred line)	40	37.5
73 (autosterile inbred line)	137	65.7
In the glasshouse:		
T57	32	9.4

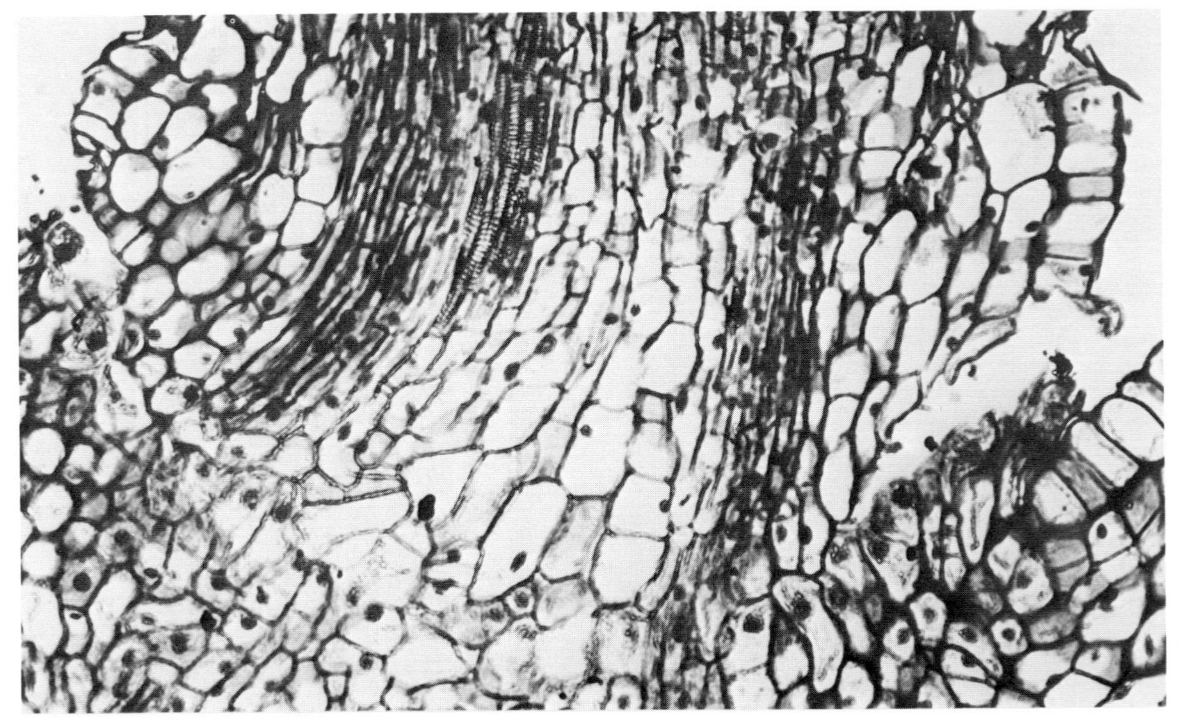

Fig. 4. (a) Separation of cortical cells at the peduncle/pedicel junction shortly before flower shedding. Stained with toluidine blue. x 60.

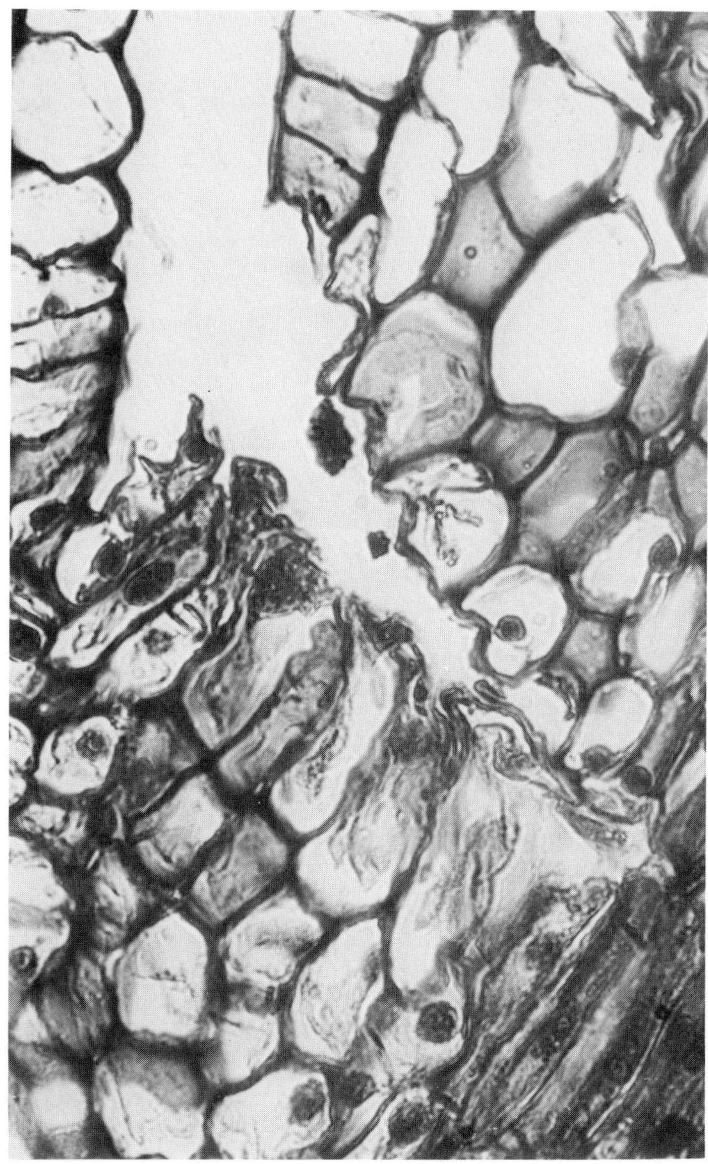

Fig. 4. (b) Lysis of the cell middle lamella at the peduncle/pedicel junction. x 150.

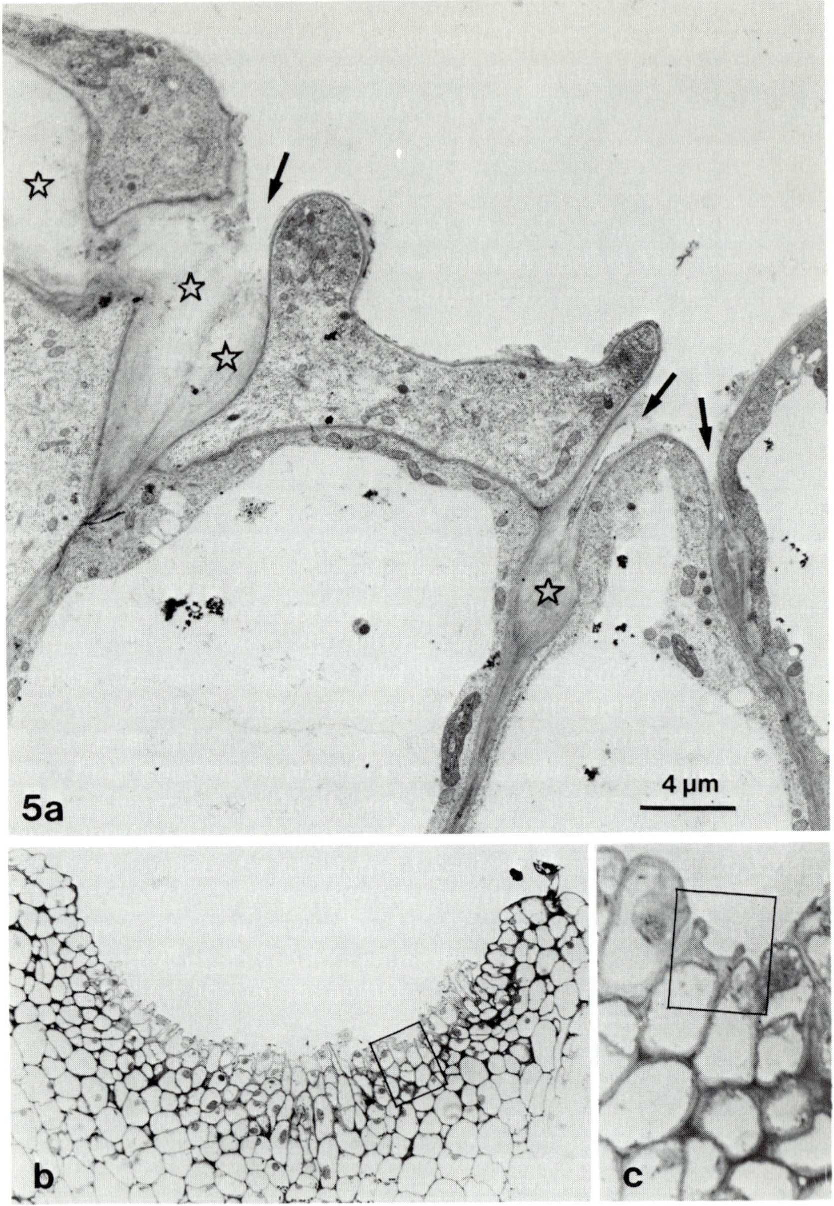

Fig. 5. Exposed peduncle cells at flower drop. (a) Electron micrograph showing separation of pedicel and peduncle has occurred along cell wall middle lamella. Some swelling (starred) of cells walls and also some intracellular separation (arrowed) between cells of exposed layer. (b) Light micrographs for location of (a) in transverse and (c) section of peduncle.

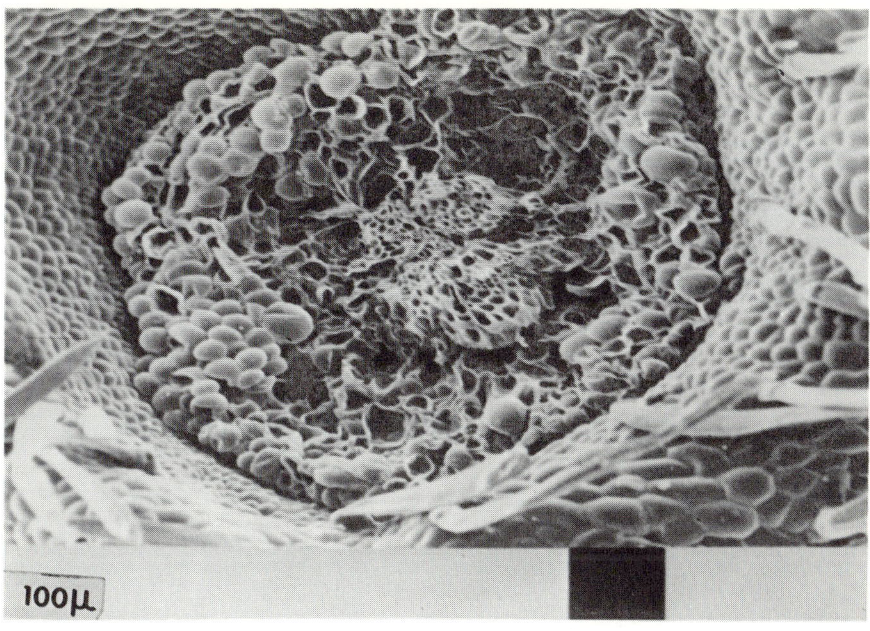

Fig. 6. (a) Scanning electron micrograph of the abscission region on the peduncle immediately after flower shedding, showing a ring of inflated cortical cells and fractured vascular bundles.

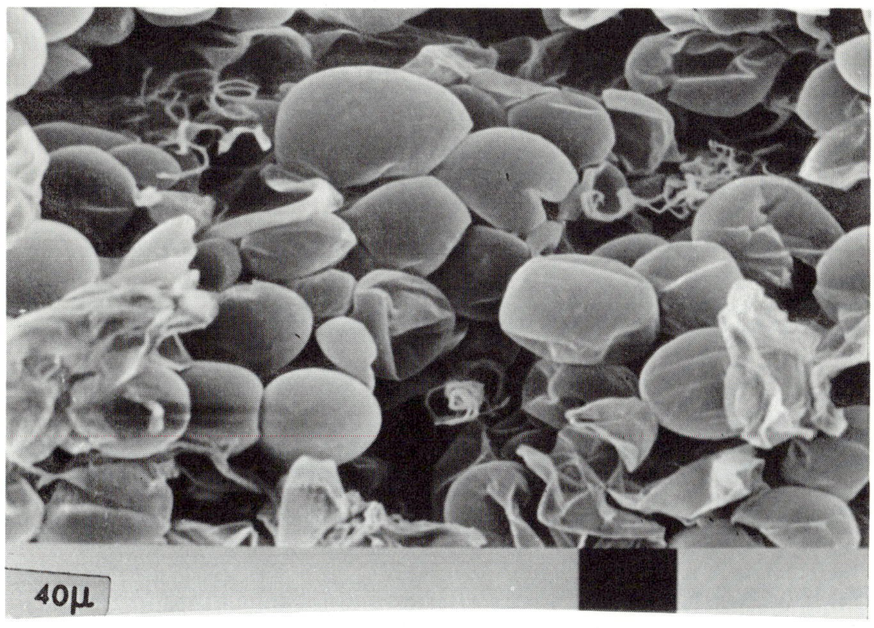

Fig. 6. (b) Inflated cortical cells of the abscission zone, with ruptured xylem spiral thickening, after flower shedding.

DISCUSSION

Failure of fertilisation has been eliminated as a major contributory factor to high levels of flower shedding in *Glycine max* (Abernathy et al., 1977) and both Kambal (1969) and Chapman et al. (1979) have demonstrated that many abscised *V. faba* flowers are fertilised. The proportion of abscised flowers with pollen tubes in the ovary (Table 1) confirm that fertilisation failure is not the sole cause of flower shedding.

The pattern of flower shedding in axillary racemes of *V. faba* and in those of other economically important legumes such as *Phaseolus vulgaris* (Subhadrabandu et al., 1978; Tamas et al., 1979) and *Vigna unguiculata* (Ojehomon, 1972) is quite characteristic, with the earliest-opening basal flowers most frequently forming pods, unlike those at the raceme apex, which are almost invariably shed (Jaquiéry and Keller, 1978).

We have developed a hypothesis to account for the observed pattern of flower drop within axillary racemes, based on our observations of flower shedding. In Northern European *V. faba*, with numerous, sequentially-opening flowers in each axillary inflorescence, the pollination of the earliest flower to reach anthesis promotes physiological changes which stimulate pod set and cause structural alterations in cell walls. These structural alterations occur both in the fertilised flower, where such activity would be required for cell expansion in young fruit development, and also in young unpollinated buds higher up the racemes. Premature lysis of the cell wall middle lamellae in the buds ensures that they are destined to abscise, even though they may open normally and be fertilised before this occurs.

Some evidence for this hypothesis comes from results in the literature as well as from our unpublished results. Pollination is known to trigger changes that promote flower wilting (Burg and Dijkman, 1969; Gilissen, 1976, 1977) or abscission of floral structures (Stead and Moore, 1979). If

V. faba stigmas are pollinated, and the styles excised 60 min. later, before the pollen tubes can reach the ovary, parthenocarpic pod set results, and furthermore ^{14}C label assimilated by basal flowers of an axillary raceme rapidly appears in buds at the raceme apex (data not presented). These results indicate that pollen tube growth releases a stimulus promoting pod set, and that such a stimulus might be transported from basal flowers to apical buds.

Our preliminary observations have shown that lines with only 4 - 5 flowers/node opening over 1 - 2 days show very little flower shedding, probably because fertilisation occurred in apical flowers before the hormonal stimulus from recently fertilised basal flowers could trigger the abscission process. In contrast, Northern European lines, with 8 - 9 flowers/node opening over a period of 5 - 6 days commonly exhibit 50 - 60% flower shedding.

Synchrony of anthesis within axillary nodes varies between genotypes, and it is suggested that by selecting for this character flower shedding might be reduced and yield stability improved.

ACKNOWLEDGEMENTS

The authors would like to thank Professor D. Boulter for many helpful discussions, and Dr. D.A. Bond of the Plant Breeding Institute, Cambridge, for the supply of seeds. Financial support from the EEC for this project is gratefully acknowledged. M.L. Smith gratefully acknowledges an SRC CASE studentship.

REFERENCES

Abernathy, R.H., Palmer, R.G., Shibles, R. and Anderson, I.C., 1977. Histological observations on abscising and retaining soybean flowers. Can. J. Plant Sci. 57, 713-716.

Burg, S.P. and Dijkman, M.J., 1967. Ethylene and Auxin participation in pollen induced fading of vanda orchid blossoms. Plant Physiol. 42, 1648-1650.

Chapman, G.P., Fagg, C.W. and Peat, W.E., 1979. Parthenocarpy and Internal competition in *Vicia faba* L. Z. Pflanzenphysiol 94, 247-255.

Gilissen, L.J.W., 1976. The role of the style as a sense organ in relation to the wilting of the flower. Planta 131, 201-202.

Gilisson, L.J.W., 1977. Style controlled wilting of the flower. Planta 133, 275-280.

Jaquiéry, R. and Keller, E.R., 1978a. Beeinflussung des Fruchtansatzes bei der Ackerbohne (*Vicia faba* L.) durch die verteilung der assimilate. Teil 1. Angew. Botanik 52, 261-276.

Jaquiéry, R. and Keller, E.R., 1978b. La chute des fruits chez la feverole (*Vicia faba* L.) en relation avec la disponibilité en assimilats marqués au ^{14}C. Revue suisse Agric. 10 (4), 123-127.

Kambal, A.E., 1969. Flower drop and fruit set in field beans, *Vicia faba* L. J. Agric. Sci., Camb. 72, 131-138.

Martin, F.W., 1959. Staining and observing pollen tubes in the style by means of fluorescence. Stain Technol. 34, 125-128.

Ojehomon, O.O., 1972. Fruit abscission in cowpea *Vigna unguiculata* (L.) Walp. I. Distribution of ^{14}C-assimilates in the inflorescence, and comparative growth of ovaries from persisting and abscising open flowers. J. Exp. Bot. 23 (76), 751-761.

Sexton, R., 1976. Some ultrastructural observations on the nature of foliar abscission in *Impatiens sultani*. Planta 128, 49-58.

Stead, A.D. and Moore, K.G., 1979. Studies on flower longevity in Digitalis: Pollen-induced corolla abscission in Digitalis flowers. Planta 146, 409-414.

Subhadrabandu, S., Adams, M.W. and Reikosky, D.A., 1978. Abscission of flowers and fruits in *Phaseolus vulgaris* L. I. Cultivar differences in flowering pattern and abscission. Crop Science 18 (5), 893-896.

Tamas, I.A., Wallace, D.H., Ludford, P.M. and Ozbun, J.L., 1979. Effects of older fruits on abortion and abscisic acid concentration of younger fruits in *Phaseolus vulgaris* L. Plant Physiol. 64, 620-622.

DISCUSSION

G.P. Chapman *(UK)*

I have a question for Dr. Gates. Can he relate the result with simultaneous flowering to that of pod development, does it follow that all the pods on the node would develop if the plant were decapitated? We have found that decaptitation can induce all the pods on a particular node to develop. I also have some data which concerns the effect of TIBA on the apex and other growth regulators on various parts of the inflorescence.

P. Gates *(UK)*

The effect of decapitation can partially upset the hormonal balance of the plant. One of the critical things in the development of the pod and in flower shedding is the overall hormone balance. Anything which disrupts that is going to affect the reproductive performance of the plant. Having said that, our group is not involved in hormones at all and we know very little about the physiology of them. We have looked at morphology and at some of the enzymes which affect the cell wall but we have not looked at any hormones and I cannot speak with any authority on the subject.

G.C. Hawtin *(Syria)*

When we look at the problem of flower drop there are two different issues. There is the flower drop within a node, as we have heard from Dr. Gates, in which hormones may play an important role, but often there is a large number of completely empty nodes. From a breeding point of view, if the aim is a 6 t/ha crop, then with a 100 seed weight around 60 g, about 1000 seeds/m^2 is required. At a planting density of around 30 - 40 plants/m^2 it means we need about 30 seeds/plant. In turn this means we need only about 6 pods per plant to get the 6 t crop. Should all 6 pods be produced at a single node, or is it better to have 1 - 2 pods at each node but spread over more nodes? Has anyone looked at the hormonal effects between nodes?

P. Gates

The situation within a node is quite complicated and we have not complicated it further by adding another layer of complexity. The only comment I can make on high sowing rates is that it might depend on the initial cost of the seed. If you plant at very high densities it may make actual seed growing very expensive.

D.A. Lawes *(UK)*

As a breeder I feel that the best crop would be one that has 6 pods per plant with 3 on each of two nodes rather than 6 on an individual node. Could a physiologist tell us whether this is a reasonable proposition? Can 3 or 4 pods on two nodes be better supported on a plant than 6 or 8 pods on an individual node?

M.H. Poulsen *(Denmark)*

I would like to add to the last question and also help the physiologists by supplying part of the answer. In the *V. faba* major type the yield develops within 3 - 4 inflorescences. To elaborate on the question I would ask whether the level of yield in broad beans could be introduced into types preferred by many farmers - that is the *V. faba* minor type of crop, but with more pods and small seeds and therefore, more seeds, whilst maintaining the yield level of broad beans.

R. Thompson *(UK)*

There is evidence that a canopy can support a higher pod load than it normally carries. I showed earlier that if you allow many more pods to set by removing competition for light during a short period of time of only 15 days, when a normal level of competition is re-imposed the assimilate produced by the canopy of normal leaf area index, will support the additional pod load. Therefore, if pod drop could be reduced, it seems that the canopy would stand a very good chance of being able to support growth of the additional seed.

J. Picard *(France)*

Can we hear the point of view of a breeder?

E. von Kittlitz *(FRG)*

In crosses between major and minor types in progenies F_3 and F_4 etc., I find a concentration of pods in the lower parts, but so far I have not found a good distribution of the seed further up the stem. I would like to ask what can really be expected from crosses between major and minor types? We hope to do some experiments in Hohenheim on this reciprocal process.

G.P. Chapman

As far as putting pods on particular nodes is concerned, one of the things which we noticed about the line I called N1 this morning, was that if you planted it at very high densities it did not grow much over ½ m tall. However, at such high densities the proportion of fertile branches was increased though the total number of branches on the plant was reduced. Also, the functional pods are formed more towards the top of the plant. There are still some on the lower nodes but most of the pods are near the top of the plant. Therefore, with that particular plant model, there is some flexibility depending on the planting density that is chosen.

J. Picard

There is a large diversity in beans. The problem is to recombine the different types. This is not impossible for the plant breeder providing there is not a strong linkage between the characters to be recombined. If there is a strong linkage then other methods must be employed, for example, through mutation. It may be that an exchange of information between physiologists and plant breeders could stimulate this kind of work.

BEE VISITS, CROSSING FREQUENCIES AND YIELD STRUCTURE OF FIELD BEANS *(Vicia faba)* WITH CLOSED FLOWER CHARACTER

J. Chr. Nørgaard Knudsen[1] and M. H. Poulsen[2]
[1]Royal Veterinary and Agricultural University, Copenhagen
[2]Danish Plant Breeding Ltd., Store-Heddinge,
Denmark

ABSTRACT

The frequency and foraging behaviour of honeybees in field beans with the closed character and in recommended varieties with normal flowers were investigated in both pure stands and in fields with alternating plots of field beans with the two flower types in 1978 and 1979. The mean number of honeybees per unit area was three to four times higher in fields and plots with normal flowered plants than in those with closed flowered. Pure stands of closed flowered field beans were only visited by pollen collecting honeybees but in 1979 closed flowered plants in alternating plots were visited by nectar collectors, although these plants do not contain any nectar.

The frequency of outcrossing in the years 1977 to 1979 were investigated on the basis of different markers. Under Danish conditions the percentage outcrossing was halved to 25% due to the introduction of the closed flower gene.

The data for phenotypic correlations and variation in seed yield components suggested that selected families with the closed flower character in field trials in 1979 suffered from an inadequate number of pollinating insects. This was taken to be evidence for a generally insufficient level of autofertility in these families. However, certain families showed seed yields comparable to those of recommended varieties and it was concluded that the closed flower character will be a useful tool in largescale selection for autofertility.

INTRODUCTION

The field bean is a naturally entomophilous species that normally requires tripping of flowers for pollination. Once flowers are tripped either self- or cross fertilisation may occur. The tripping requirement may lead to instability in yield due to inadequate numbers of honey- and bumble-bees especially in large fields and under unfavourable climatic conditions. With the aim of increasing yield reliability, genotypes with the ability to self-pollinate and produce seed in the absence of flower visiting insects are sought (Lawes, 1974). Types with a concentrated seed set in the first inflorescences are considered essential in combining high yields and earliness in ripening (Poulsen, 1977b).

Poulsen (1977a) reported on the closed flower mutant where the flower opens only approximately one third of that observed in normal flowers. In addition the mutant flowers are without nectar and scent. The closed flower was shown to be inherited by a single recessive gene designated cf. It was suggested that this new flower type might be a useful tool in breeding autogamous field beans. This paper presents preliminary results from studies on bee activity, crossing frequencies and yield structure of field beans with the closed flower character.

MATERIALS AND METHODS

Bee visits

The frequencies and foraging behaviour of honey- and bumble-bees were investigated in fields with pure stands of field beans with either normal or closed flowers in 1978 and 1979. In 1978 the two fields were located in opposite directions from the apiary at a distance of approximately 300 m. In 1979 observations were taken in two nurseries each planted with lines of a single flower type. The two nurseries were located next to each other at a distance of 400 m from the apiary. In both years observations of bee frequencies and foraging behaviour were also taken in fields with alternating plots of field beans with the two flower types.

Observations were taken at 13 h, 15 h and 17 h (GMT+1) on six days in 1978 and 9 days in 1979. On each occasion bees were counted in five 10 m^2 plots of each flower type in pure stands and in fields with alternating plots. The mean bee numbers calculated for each observation-time and day were transformed to number of bees per hectare.

Outcrossing

Crossing frequencies have been estimated on the basis of different characters in different years. In 1977 a population segregating for the closed flower character were grown at PBI Cambridge. Single plants with the closed flower character were selected and the progeny grown in 1978 at Højbakkegård, Tåstrup. The crossing frequency was estimated by counting the number of normal flowered plants and multiplying by 1.5 since half of the crossings with heterozygous plants were not reflected in the field. In 1978 the lines were grown in a nursery. Two families with the topless character (Sjödin, 1971) were selected and the crossing frequency was estimated in 1979 on the basis of normal indeterminate plants in the progeny. Among families in the nursery in 1979 a few white flowered plants occurred. These plants were selected and the seeds were sown in 1980 and the crossing frequencies estimated on the basis of the proportion of plants with coloured flowers among the progeny.

Yield trial

Twentyeight families with the closed flower character were included in field trials in 1979 together with 20 normal flowered varieties. The families originated from two generations of pedigree selection. Sister lines were bulked after rogueing atypical plants considered to be crosses. The families were thus expected to represent two to three generations of inbreeding by selfing.

The experiment was laid out as a randomised block design with four replicates and 50 varieties. Plot size was 7.5 m^2 with a seeding rate of 45 seeds per m^2. During the growing season observations were taken on flowering, ripening and lodging.

Shortly before harvest five plants were taken from each plot and analysed for their complete yield structure. The remainder of the plot was harvested as a seed yield trial. Phenotypic correlations between seed yield components were computed from the mean values of the five plants from the individual plots. In a number of plots plant establishment was poor and plots with less than 25 plants per m^2 were omitted when calculating the correlations.

RESULTS

The results from observations on honeybee visits and their foraging behaviour in normal and closed flowered field beans are given in Table 1. In fields with pure stands of closed flowered beans the mean number of honeybees per ha was reduced to 25% of the number found in fields with normal flowered plants.

TABLE 1

MEAN NUMBER OF HONEYBEES PER ha OF FIELD BEANS OF NORMAL AND CLOSED FLOWER TYPES AND THE PROPORTION OF POSITIVE VISITS IN FIELDS WITH PURE STANDS AND IN FIELDS WITH ALTERNATING PLOTS OF THE TWO FLOWER TYPES

Year	Pure stands				Alternating plots			
	Normal flower type		Closed flower type		Normal flower type		Closed flower type	
	Mean No.	% pos. visits	Mean No.	% pos. visits	Mean No.	% pos. visits	Mean No.	% pos. visits
1978	2080	94.7	570	100	2190	95.5	540	100
1979	3290	79.5	670	100	1870	88.3	550	87.7

The reduction in mean number of bees per ha when comparing normal and closed flowered plants in alternating plots was of the same magnitude as in pure stands in 1978, whereas in 1979 the reduction was slightly less. In pure stands of closed flowered field beans all the visiting bees were pollen collectors referred to in the Table as positive visits. In alternating plots different results were obtained in 1978 and in 1979. In 1978 honeybees were pollen collectors but in 1979 12.3% of the visiting bees were nectar collectors trying to collect nectar from closed flower field beans despite the fact that these do not contain any.

Bumblebees were only occasionally observed in fields or plots with closed-flower plants. The species *Bombus hortorum* and *B. lapidarius* were observed as pollen collectors in both flower types. The dominant species *B. terrestris* was common in normal flowered types where it robbed nectar through a hole bitten in the corolla tube. On average 79% of the normal flowers had their corolla tube perforated whereas only 1.2% of the closed flowers had been bitten by bumblebees. The observed crossing frequencies in three successive years are given in Table 2.

TABLE 2

CROSSING FREQUENCIES IN FIELD BEANS WITH THE CLOSED FLOWER CHARACTER ESTIMATED IN THREE SUCCESSIVE YEARS ON BASIS OF DIFFERENT MORPHOLOGICAL CHARACTERS

Year	Location	Character	Mean	Range	Std. dev.
1977	Cambridge	Flower type	6.1	0.4 - 40.3	15.01
1978	Tåstrup	Determinate growth	20.1	19.0 - 21.1	-
1979	Tåstrup	White flowers	22.9	0.0 - 60.4	15.83

The material grown in the field at PBI Cambridge showed a crossing frequency of 6.1% covering a range from 0.4% to 40.3%. The lowest values observed were in lines originating from crosses with autofertile African material. The outcrossing in the two topless lines in 1978 averaged 20.1% and in 1979 the outcrossing in 40 white-flowered lines averaged 22.9% with a range from zero to 60.4%.

In Table 3 the mean values, ranges and standard deviations for a number of seed yield components are given for 20 normal flowered commercial varieties and for 28 families with the closed flower character. Within both groups a great variation was present but no systematic differences due to flower type was evident from the data.

The phenotypic correlations between seed yield components are given separately for closed- and normal flowered lines in Table 4.

TABLE 3

MEAN, RANGE AND STANDARD DEVIATION FOR YIELD AND YIELD COMPONENTS IN 20 NORMAL-FLOWERED VARIETIES AND 28 CLOSED FLOWERED LINES OF FIELD BEANS GROWN IN THE FIELD IN 1979.

Component	Normal flowers			Closed flowers		
	Mean	Range	Std. dev.	Mean	Range	Std. dev.
Plot yield (hkg/ha)	41.4	29.2 - 53.2	5.77	34.6	19.8 - 50.2	5.24
Yield per plant (g)	23.2	12.8 - 38.9	4.65	22.9	9.6 - 43.9	5.60
No. of podded nodes	9.3	6 - 13	1.89	10.0	4.0 - 15.0	2.31
No. of pods	18.3	12.0 - 32.2	3.55	21.6	9.4 - 45.4	5.53
Pods per podded node	2.0	1.1 - 3.7	0.47	2.2	1.5 - 3.2	0.40
No. of seeds	59.7	36.9 - 89.9	10.91	55.5	19.6 -133.8	17.28
Seeds per pod	3.3	2.6 - 4.0	0.25	2.6	1.9 - 3.2	0.29
Seed weight (mg)	396	249 - 640	80.0	425	295 - 580	53.1

TABLE 4

PHENOTYPIC CORRELATIONS BETWEEN YIELD COMPONENTS. UPPER MATRIX FROM 13 FAMILIES WITH THE CLOSED-FLOWER CHARACTER AND LOWER MATRIX FROM 20 NORMAL-FLOWERED VARIETIES

Component	1	2	3	4	5	6	7
1. Number of pod bearing nodes	-	.82**	.69*	.40	-.63*	-.13	-.44
2. Number of pods	.17	-	.83***	.80**	-.07	-.18	-.17
3. Number of seeds	.12	.92***	-	.73**	-.05	.39	-.51
4. Seed yield/plant	.18	.26	.39	-	.37	-.07	.22
5. Pods/bearing node	-.67**	.61**	.59**	.06	-	.02	.50
6. Seeds/pod	-.09	-.02	.36	.44*	-.05	-	-.65*
7. Seed weight	.07	-.57**	-.52*	.58**	-.47*	.09	-

*, ** and *** indicates .05, .01 and .001%

Seed yield per plant is positively related to all the seed yield components except to seeds per pod in the closed flower group. In both groups a significant negative correlation was found between the number of pod-bearing nodes and the number of pods per podded node. In the closed flower group the number of pods and consequently the number of seeds were dependent on the number of pod-bearing nodes whereas this was not the case among the commercial varieties. In this latter group the number of pods per podded node contributed significantly to the total number of pods and seeds. In both groups seeds per pod was positively related to the total number of seeds. Negative correlations of seed weight to the number of pods and number of seeds were found in both groups. Among the closed flower families the number of pod-bearing nodes was negatively correlated to seed weight presumably as a consequence of the positive relationship between earliness and seed weight ($r = 0.20$). This last relationship was negative in the commercial varieties. The two groups also show differences in the relationships of pod-bearing nodes to seeds per pod and seed weight. Among the closed flower families, seed weight and the number of seeds per pod were negatively correlated and the latter yield component thus did not contribute much to the total seed yield. A positive correlation was found between the number of pods per podded node and seed weight. Among the normal flowered commercial varieties pods per podded node show a negative correlation with seed weight but no relationship was found between seeds per pod and seed weight. Three closed flower lines are compared to commercial varieties in Table 5. The best closed flowered lines showed a seed yield performance similar to commercial varieties of the same earliness.

DISCUSSION

The differences in number of bees in pure stands in both years and in alternating plots in 1978 was partly accounted for by the total absence of nectar-collecting bees in the closed flower material. In 1979 a higher proportion of the honeybees observed were nectar collectors. In alternating plots these tried to collect nectar from closed flowers probably 'misled' by the mixed plots since attraction by scent to the closed flowers does not take place.

On the basis of the pollen-collecting bees the number of visiting bees was still reduced to about one third in the closed flower material compared to normal flowered. This avoidance may be due to increased difficulties for the bees in entering the flowers for pollen as a consequence of the reduced opening of the flower.

Bumblebees were seldom observed in fields and plots with closed flower material and these constituted only a small proportion of the observed pollinating insects. It can thus be expected that if field beans with the closed flower character are placed in pure stands in locations a great distance from apiaries the number of pollinating insects per unit area will be drastically reduced and the possibilities for large-scale selection of genotypes with improved self-fertility consequently increased.

The observed crossing frequency in material grown at Cambridge in 1977 (6%) is lower than those observed in material grown at Tåstrup, in 1978 and 1979 (20 and 23% respectively). Poulsen (1977a) reported a crossing frequency of 6% in material grown in Cambridge in 1975 and 1976 and the available data points to differences between the two locations. In normal flowered field beans, Poulsen (1971) reported a crossing frequency of approximately 50% at locations in Denmark where conditions were favourable for insect pollination. The conditions for insect pollination of the closed flower material in 1978 and 1979 may be argued to have been favourable due to the high number of honeybees per unit area (Table 1).

The introduction of the closed flower character has thus halved the frequency of outcrossing under the conditions prevailing at the experimental farm Højbakkegård, Tåstrup.

TABLE 5

COMPARISON OF EARLINESS, YIELD AND YIELD COMPONENTS OF THREE SELECTED CLOSED FLOWERED FAMILIES AND A NUMBER OF STANDARD VARIETIES OF FIELD BEANS IN FIELD TRIALS IN 1979

Variety	Ripening (week no.)	Plot yield (hkg/ha)	Plant yield (g/plant)	Seedweight (mg)	No. of pods	No. of seeds
C45 (cf,cf)	36.75^{a}	37.1^{ab}	19.9^{ab}	441^{c}	18.8^{abc}	45^{cd}
Diana	36.75^{a}	41.5^{bc}	20.7^{ab}	328^{ab}	19.6^{ab}	62^{abc}
Sving	36.75^{a}	33.8^{a}	20.7^{ab}	277^{a}	20.8^{a}	75^{a}
Herz freya	36.75^{a}	36.7^{ab}	23.1^{ab}	416^{cd}	17.2^{abc}	56^{bcd}
C46 (cf,cf)	37^{a}	39.9^{b}	19.0^{b}	445^{c}	18.0^{abc}	42^{d}
Ascott	37.5^{a}	42.4^{c}	22.0^{b}	425^{d}	16.3^{bc}	52^{bcd}
P27 (cf,cf)	38.75^{b}	41.3^{bc}	24.7^{a}	456^{d}	20.3^{ab}	54^{bcd}
Maris Bead	39^{bc}	37.2^{ab}	21.4^{ab}	354^{bc}	18.5^{abc}	60^{b}
Blaze	39.5^{bc}	44.6^{c}	24.5^{a}	489^{d}	15.4^{c}	51^{bcd}

Figures with the same letter are not significantly different at the 5% probability level

The wide variation in outcrossing found among lines especially in 1979 may be explained in differences in the degree of autofertility. However, as also mentioned by Poulsen (1977a), pronounced differences in flower opening are found between lines with the closed flower gene and these differences may have influenced insect pollination and subsequently the observed outcrossing.

The values for seed yield components given in Table 3 show no pronounced difference between the two groups. Two to three generations of pedigree selection with increasing inbreeding in the closed flower material has not in general led to inbreeding depression.

The present studies indicate that tripping and pollination in the closed flower material has been insufficient for a good seed set in lower nodes and on the other hand that in general the degree of autofertility has not been raised to a level that compensates for the reduced insect visits.

The negative relationship between number of seeds per pod and seed weight found among closed flower families probably relates back to the parents from some of the initial crosses including African material with a small number of relatively large seeds per pod.

Despite the fact that sufficient levels of autofertility have not been introduced into the families, a number of selections (Table 5) have shown promising results with seed yields comparable to those of commercial varieties. These results are obtained after three generations of selection for both agronomic type and autofertility and indicate that it is possible by means of the closed flower character to select lines with increased self-fertility and thus achieve a stabilising effect on yield.

REFERENCES

Lawes, D.A. 1974. Field beans: improving yield and reliability. Span 17 (1)

Poulsen, M.H. 1971. Honningbiernes (*Apis* L.) og humlebiernes (*Bombus Latr.*) arbejdsmåde og betydning for frøsaetningen i hestebønne (*Vicia faba* L.) PhD thesis, Royal Veterinary and Agricultural University, Copenhagen, Denmark.

Poulsen, M.H. 1977a. Obligate autogamy in *Vicia faba* L. J. agric. Sci., Camb. 88:253-256

Poulsen, M.H. 1977b. Genetic relationships between seed yield components and earliness in *Vicia faba* L. and the breeding implications. J. agric. Sci. Camb. 89:643-654

Sjödin, J. 1971. Induced morphological variation in *Vicia faba* L. Hereditas 67:155-180

FACTORS INFLUENCING FERTILITY OF A TETRAPLOID IN *Vicia faba*

A. Martín and J.A. Gonzalez-García

Departamento de Genética,
Escuela Técnica Superior de Ingenieros Agrónomos,
Córdoba, Spain.

ABSTRACT

Meiotic behaviour and fertility of an autotetraploid Vicia faba *are described in relation to effects of temperature and agronomic practice.*

Positive correlations existed between number of quadrivalents and fertility. Sowing date was an important determinant of fertility but removal of the shoot apex did not improve seed set. The regularity of meiosis as affected by temperature was assessed by counting the number of micronuclei at the tetrad stage.

INTRODUCTION

Vicia faba has been used extensively in cytological studies. However, despite many attempts, there are no well substantiated records of success in obtaining fertile aneuploids or polyploids until the recently reported tetraploids of Poulsen and Martín (1977) and trisomics of Martín (1978).

Polyploidy has both physiological and genetic consequences; this has been recognised since the early work with induced autopolyploidy by Muntzing (1936).

The increased size characteristic of autopolyploids is not found in the *V. faba* tetraploid; only the seeds, flowers and leaves are slightly larger than those of cv. Primus, the putative variety from which the tetraploid was produced. There are, however, two characters which are typical of this material, which are the wavy aspect of the leaf edges and the pollen shape, identical to that reported by Sjödin (1971) as a pollen mutant (PO-1). The tetraploid has, on the other hand, agronomic disadvantages compared with the diploid, namely lateness, both in development and in ripening, low level of resistance to diseases and to adverse environmental conditions, lower tillering and generally very low fertility.

The sterility of autotetraploid plants is generally attributed to three phenomena (Stebbins, 1947):

1. Irregular chromosome distribution caused by unequal separation of multivalents.
2. Irregular distribution caused by meiotic abnormalities of a physiological nature.
3. Genetic-physiological sterility not associated with meiotic irregularity.

There is contradictory experimental evidence of the existence both of a heritable component of variation controlling the pattern of chromosome pairing and whether such variation influences fertility. Fertility related to increase in quadrivalent number has been reported in rye (Muntzing, 1951; Hazarika and Rees, 1967), barley (Tsuchiya, 1957) and *Dactylis* (McCollum, 1958), while the opposite has been found in maize (Gilles and Randlof, 1951), *Brassica campestris* (Swaminathan and Sulbha, 1959) and barley (Bender and Gaul, 1966). Other authors have found fertility related to increase in bivalent number (Bremer and Bremer-Reinders, 1954; Hilpert, 1957) while Morrison and Rajhathy (1960) did not find any link.

Both environment and genotype influence the pairing pattern. Thus, in rye, (Muntzing, 1951) it was found that quadrivalent formation is determined genetically if a response can be shown to selection.

In relation to environmental conditions, many studies since Darlington (1940) have confirmed the deleterious effect of high temperature. Pao and Li (1948) observed that high temperature reduces quadrivalent formation in rye. Hossain (1978) reported the same effect but found stable bivalent frequency. High and low temperatures decrease pairing on hexaploid wheat, this being under genetic control (Riley, 1966; Riley et al., 1966), and Lu (1974) stated that high temperature increases recombination in *Coprinus*.

Other investigations have demonstrated the effects of ionic environment and mineral nutrition on chromosome pairing in different species (Steffensen, 1957; Law, 1963; Bennet and Rees, 1970; Fedak, 1973).

The aim of this work was to study meiotic behaviour in relation to reproductive fertility, the effect of temperature on meiosis and the effect of agronomic practices on the fertility of the *V. faba* autotetraploid. The reason for this was to obtain information to be used in a long-term programme directed at the diploidisation of the autotetraploid faba bean.

MATERIAL AND METHODS

The origin of the tetraploid line has been discussed in a previous paper (Poulsen and Martín, 1977).

The effect of temperature was studied on plants grown in a controlled environment chamber. Initially, the growing conditions were an 18 h day at 22°C and a 6 h night at 18°C. After meiosis the growing conditions were changed to a day temperature of 12°C and night temperature of 8°C without changing daylength, and again flowers were taken for meiotic analysis.

Observations on meiotic behaviour were made (by only one person) on squash preparations of anthers fixed in 1 : 3 aceto-orcein. It was very difficult to obtain plates with all the chromosomes spread out. For this reason only the bivalents and univalents clearly visible were counted.

Counts of the somatic chromosome number were made from squashes of root-tips pre-heated in a 0.05% colchicine solution, fixed on 1 : 3 acetic alcohol and stained by the Feulgen procedure.

Pollen fertility data on 47 plants were obtained from slide preparations stained with aceto-carmine.

Plants were decapitated at the mid-point of the eleventh node as early as possible.

Estimates of the numbers of flowers/node were made from the average of the flowers on the first ten nodes of the main branch. Total numbers of flowers were calculated by multiplying numbers of flowers/node by 10 and by the number of branches in which there was pod set. Plants were grown in insect-proof cages.

RESULTS

Selection and aneuploidy

In the last two years, one of the tetraploid lines (number 45) showed greater fertility than the rest of the tetraploid material studied.

Tables 2 and 3 show the data from two sowing dates, in which the relatively more fertile line 45 was compared to a control. At the second sowing date there were no significant differences, but, at the first, some fertility characters and the first flowering node differed significantly between the two lines. These differences were due to the lower fertility of the control as the other line showed no change in behaviour.

The study of aneuploidy in the progeny of 24-chromosome plants was hindered by the morphology of *V. faba* chromosomes (5 of them are unidentifiable). It cannot be assumed that a 24-chromosome plant is aneuploid (it could be trisomic for one chromosome and pentasomic for another, for example).

For the control line, of 36 counts 28 plants had 24 chromosomes, 7 had 25 and 1 had 23, i.e. 22.2% of aneuploidy. In line 45, out of 25 plants, 19 had 24 and 6 had 25 chromosomes, i.e. 24% of aneuploidy. This did not represent a significantly higher frequency of univalents than with the 24-chromosome plants.

Table 1 shows the data of the comparison between 24- and 25-chromosome plants. Twenty-four-chromosome plants produced more grains/plant and more pods/plant than the 25-chromosome plants.

Meiotic behaviour

The effect of temperature on the meiotic behaviour can be seen in Tables 4 and 5. The only clear result was the significant effect of temperature on micronuclei at the tetrad stage.

TABLE 1

SOME REPRODUCTIVE AND VEGETATIVE CHARACTERS OF 4X and 4X+1 *Vicia faba* L. ($\bar{X} \pm S_{\bar{X}}$)

Chromosome number (1)	Seeds/plant	Pods/plant	Branches/plant	Flower/node	First node with flower
24 n = 44	7.84 ± 1.30	6.20 ± 1.00	2.88 ± 0.23	7.53 ± 0.15	10.7 ± 0.35
25 n = 13	1.23 ± 0.39	1.23 ± 0.39	2 ± 0.27	7.01 ± 0.40	10.69 ± 0.47
t values (2)	2.7**	2.3*	1.9 NS	1.4 NS	0.02 NS

(1) n = number of plants
(2) NS: no significant differences; * $0.05 > P > 0.01$; ** $P < 0.01$

TABLE 2

REPRODUCTIVE AND VEGETATIVE CHARACTERS IN LINES SELECTED FOR HIGHER FERTILITY AND UNSELECTED LINES (DATE OF SOWING, 5-12-79) ($\bar{X} \pm S_{\bar{X}}$)

Line (1)	Seeds/plant	Pods/plant	Branches/plant	Flowers/node	Pods/flowers	First node with flower
Highly fertile 4X n = 17	16.47 ± 3.03	11.94 ± 2.08	2.71 ± 0.33	7.68 ± 0.27	0.09 ± 0.02	9.58 ± 0.34
Normal population 4X n = 29	6.45 ± 0.81	5.41 ± 0.77	3.24 ± 0.29	7.29 ± 0.16	0.05 ± 0.01	11.17 ± 0.40
(2) t values	3.76**	3.40**	1.15 NS	1.31 NS	2.20*	2.78**

(1) n = number of plants
(2) NS: no significant differences; * $0.05 > P > 0.01$; ** $P < 0.01$

TABLE 3

REPRODUCTIVE AND VEGETATIVE CHARACTERS IN LINES SELECTED FOR HIGHER FERTILITY AND UNSELECTED LINE (DATE OF SOWING 27-1-80) ($\bar{X} \pm S_{\bar{X}}$)

Line (1)	Seeds/plant	Pods/plant	Branches/plant	Flowers/node	Pods/flowers	First node with flower
Highly fertile 4X n = 16	15.56 ± 2.43	11.69 ± 1.79	4.5 ± 0.32	7.39 ± 0.24	0.07 ± 0.01	11.88 ± 0.32
Normal population 4X n = 19	10.74 ± 1.93	8.41 ± 1.50	4.11 ± 0.37	7.3 ± 0.31	0.06 ± 0.01	11.32 ± 0.40
t values (2)	1.52 NS	1.37 NS	0.76 NS	0.22 NS	0.72 NS	1.03 NS

(1) n = number of plants
(2) NS: no significant differences.

TABLE 4

EFFECT OF TEMPERATURE ON MEIOTIC BEHAVIOUR OF TWO LINES OF TETRAPLOID *Vicia faba*

Line		Temperature 22°C				12°C			
		Univalents	Bivalents	Micronuclei (Diads)	Micronuclei (Tetrads)	Univalents	Bivalents	Micronuclei (Diads)	Micronuclei (Tetrads)
45	$\bar{X} \pm S_{\bar{X}}$ No. of plants	19.5±3.01	21 ±4.13	11.6±1.49	14 ±1.08	9.2±1.21	34.3±5.80	10.8±0.96	8.17±0.76
29	$\bar{X} \pm S_{\bar{X}}$ No. of plants	9.3±2.14	8.3±1.73	18.4±1.49	14.6±2.83	10.3±2.6	7.7±0.55	12.6±1.17	10.71±0.69

TABLE 5

COMPARISON BETWEEN DATA GIVEN IN TABLE 4 (T VALUES) (1)

	Univalents	Bivalents	Micronuclei (Diads)	Micronuclei (Tetrads)
Line 45 at 22°C *v* Line 45 at 12°C	2.9*	1.7 NS	0.4 NS	4.0**
Line 29 at 22°C *v* Line 29 at 12°C	0.3 NS	0.2 NS	2.8*	3.1**
Line 45 at 22°C *v* Line 29 at 22°C	2.5*	2.6*	2.9*	0.4 NS
Line 45 at 12°C *v* Line 29 at 12°C	0.4 NS	3.3*	1.1 NS	2.3 NS

(1) NS: no significant differences: * $0.05 > P > 0.01$: ** $P < 0.01$

As these are a reflection of former events it is concluded that sampling error, environmental effects or other unidentified factors masked possible effects. For any given temperature, line 45 showed most bivalents.

There were significant negative correlations between both grains/plant and pods/plant with univalent and bivalent frequency (at 5% and 1% levels of probability, respectively). No correlation was found between pollen fertility and either meiotic behaviour or fertility. The number of bivalents were positively correlated (r = 0.74**). Also there was a positive correlation between univalent and bivalent numbers and micronuclei at the tetrad stage (0.53* and 0.61**).

Effect of date of sowing

Table 6 shows the results from two different dates of sowing. This is one of the most clear effects of the whole study. It is clear that late sowing increases fertility in the tetraploid.

Effect of preventing insect interference

In Table 7, data are given from plants grown in insect-proof cages and in the open. Only flowers/node and first flowering node were significantly affected. Plant height was not recorded but it was evident that plants in cages were taller than in the field, and the vegetative period was also longer. The shortening of the vegetative period for plants in the open gave rise to a notable decrease in seeds/plant.

Effect of top-removal

Data from the study on topping are presented in Tables 8 and 9. Only flowers/node and seeds/pod were significantly affected and only for one date of sowing.

Although not significant, there was a suggestion of higher fertility on undecapitated plants, whilst the opposite was found in diploid *V. faba* (Chapman et al., 1978).

TABLE 6

REPRODUCTIVE AND VEGETATIVE CHARACTERS FROM TWO SOWING DATES

Date of sowing		Seeds/plant	Pods/plant	Branches/plant	Flowers/node	First node with flower	Pollen fertility
5-12-79	No. of plants	6.11±1.00	4.77±0.72	2.61±0.16	7.29±0.12	10.58±0.22	77.22±1.35
27- 1-79	No. of plants	9.64±1.44	7.38±1.09	4.21±0.21	7.29±0.16	11.75±0.23	70.79±1.64
t values (1)		2.05*	2.06*	6.13**	0.01 NS	3.37**	2.93**

(1) NS: no significant differences: * $0.05 > P > 0.01$: ** $P < 0.01$

TABLE 7

REPRODUCTIVE AND VEGETATIVE CHARACTERS FROM TWO GROWING CONDITIONS ($\bar{X} \pm S_{\bar{X}}$)

Growing conditions (1)	Seeds/plant	Pods/plant	Branches/plant	Flowers/node	First node with flower
Open field n = 29	5.38±1.70	7.90±1.60	4.35±0.44	6.11±0.21	8.66±0.39
Insect-proof cage n = 47	9.64±1.44	7.38±1.09	4.21±0.21	7.29±0.16	11.74±0.24
t values (2)	1.86 NS	0.27 NS	0.32 NS	4.41**	7.04**

(1) n = number of plants:

(2) NS: no significant differences: * $0.05 > P > 0.01$: ** $P < 0.01$

TABLE 8

EFFECT OF TOP REMOVAL ON REPRODUCTIVE AND VEGETATIVE CHARACTERS ($\bar{x} \pm S_{\bar{x}}$) (DATE OF SOWING 5-12-79)

Plants (1)	Seeds/plant	Pods/plant	Branches/plant	Flowers/node	Pods/flowers	Seeds/pod
Decapitated n = 14	5.93±2.34	4.93±2.00	2.0±0.30	6.97±0.30	0.07±0.03	0.67±0.16
Control n = 15	13.8 ±3.57	9.6 ±2.30	2.4±0.23	8.04±0.29	0.08±0.02	1.23±0.15
t values (2)	1.75 NS	1.47 NS	0.85 NS	2.45*	0.24 NS	2.44*

(1) n = number of plants

(2) NS: no significant differences: * 0.05 > P > 0.01: ** P < 0.01

TABLE 9

EFFECT OF TOP REMOVAL ON REPRODUCTIVE AND VEGETAIVE CHARACTERS (DATE OF SOWING 27-1-80)

Plants (1)	Seeds/plant	Pods/plant	Branches/plant	Flowers/node	Pods/flowers	Seeds/pod
Decapitated n = 12	11.91±3.00	8.75±2.17	4.92±0.32	7.13±0.23	0.06±0.01	1.01±0.18
Control n = 8	13.25±3.93	10.25±2.98	3.25±0.34	7.78±0.31	0.06±0.02	1.13±0.16
t values (2)	0.26 NS	0.4 NS	1.9 NS	1.65 NS	0.0 NS	0.44 NS

(1) n = number of plants

(2) NS: no significant differences: * 0.05 > P > 0.01: ** P < 0.01

DISCUSSION

As stated in the introduction, polyploidy has complex consequences, and the experimental evidence is contradictory in relation to the effect of meiotic behaviour on fertility.

The practical impossibility of determining whether or not a plant with a chromosome number of 24 is a balanced tetraploid, due to the difficulty in distinguishing between five of the six pairs of *V. faba* chromosomes, makes study difficult. For this reason, our calculation of the frequency of aneuploids (around 20%) is a minimal estimate.

As is well-known, the performance of aneuploids is poorer than the balanced tetraploid (Sybenga, 1972) and our results are in accordance with these assumptions on fertility. We used this fact to select 20 plants with a chromosome number of 24 in which vegetative and fertility characters were not too low, and the correlation coefficients were calculated on this basis.

As in rye, quadrivalents apparently favour fertility, as fertility is negatively correlated to univalent and bivalent formation. This is difficult to explain because irregular anaphase separation often follows quadrivalent formation and results in aneuploidy in the gametes (Sybenga, 1972). The positive correlation of univalents and bivalents with irregular telophase II could mean that in tetraploid *V. faba*, as in rye, there is a marked regularity of orientation of quadrivalents.

Unfortunately, the data of the experiment with two lines at different temperatures apparently do not confirm the former results.

Line 45 was more fertile than line 29 but showed more bivalents. However, meiotic irregularity measured by micro-nuclei at the tetrad stage did not show differences. The data in Table 4 shows that the results of micronuclei analyses are the most reliable, and the levels of bivalent frequency in line 45 were the least; these are contradictory, if we accept as true the positive correlation between bivalent and micronuclei at the tetrad stage.

To reconcile these results, we must assume that line 45 is little more fertile than line 29 and that measurement of bivalent frequency of plants growing in artificial conditions is not very reliable.

The most significant effect of temperature was the increase in meiotic regularity at low temperature measured by micronuclei at the tetrad stage.

When effects of sowing date were studied, despite what was expected, late sowing produced greater seed numbers. The reason for this unexpected result may well be the sensitivity of the autotetraploid to adverse conditions and, as development is slower than in the diploid in winter, the tetraploid suffers more than the diploid. Again the results are difficult to explain because the selected fertile line was insensitive to sowing date, as is shown in Tables 2 and 3.

On the other hand, this reinforces the discussion on meiotic behaviour where we proposed that line 45 was not more fertile than line 29 in artificial conditions. For the last two years we have selected from the late sowing because we lost the first sowing due to problems such as susceptibility of the tetraploid to soil-borne fungi.

Two effects were confounded in the experiment on effects of insect visits. The net of the insect-proof cage created an environment that was particularly favourable to the development of the tetraploid allowing it a longer vegetative period than in the field. This resulted in better pod filling, as can be seen in Table 7, and therefore no conclusions can be reached.

Bennet (1971) has proposed the term 'nucleotype' to describe those conditions of nucleus that affect the phenotype independently of the informational contents of DNA. We must assume a nucleotype effect on the tetraploid response to top removal that gives rise to a response opposite to that in the diploid. From the second sowing date there were no significant differences, but the control appeared to be more fertile. On the first sowing date, the situation was quite similar but there was a significant increase in the proportion of flowers set and numbers of seeds per pod: exactly the opposite of the results obtained by Chapman and co-workers (1978) on diploid *V. faba* .

The reason could be that pod-setting is so poor that there is no competition between developing structures. Top removal may cause some damage to plants, resulting in a slightly better performance in the undecapitated ones, which also have more flowers and, therefore, the chance of better pod-set.

To conclude, it has emerged that:

1. The tetraploid faba bean seems to be very sensitive to environmental conditions, which is very common in such material.
2. It seems that there is a positive correlation between number of quadrivalents and fertility. It is evident that the latest date of sowing in our study produced the best results.
3. There was evidence that temperature affected meiotic regularity measured by the number of micronuclei at tetrads.
4. As stated in the introduction, this is the first survey performed within a long-term programme aimed at studying the possible diploidisation of the tetraploid faba bean. Many more generations of selection are required to reach definite conclusions.

REFERENCES

Bender, K. and Gaul, H., 1966. Zur Frage der Diploidisierung autotetraploider Gerste. Z. Pflang., 56, 164-183.

Bennet, M.D., 1971. The duration of meiosis. Proc. Royal. Soc. B., 178, 277-299.

Bennet, M.D. and Rees, H., 1970. Induced variation in chiasma frequency in rye in response to phosphate treatments. Genet. Res. (Cambridge) 16, 325-331.

Bremer, G. and Bremer-Reinders, D.E., 1954. Breeding of tetraploid rye in the Netherlands. I Methods and cytological investigations. Euphytica, 3, 49-56.

Chapman, G.P., Guest, H.L., Peat, W.E., 1978. Top-removal in single stem plants of *Vicia faba* L. Z. Pflanzenphysiol. 89. S. 119-127.

Darlington, C.D., 1940. The prime variable of meiosis. Biol. Rev., 15, 307-321.

Fedak, G., 1973. Increase chiasma frequency in desynaptic barley in response to phosphate treatments. Can. J. Genet. Cytol., 13, 647-649.

Gilles, A. and Randolph, L.G., 1951. Reduction of quadrivalent frequency in autotetraploid maize during a period of ten years. Am. J. Botany, 38, 12-17.

Hazarika, M.H. and Rees, H., 1967. Genotypic control of chromosome behaviour in rye. X. Chromosome pairing and fertility in autotetraploids. Heredity, 22 (3), 317-332.

Hilpert, G., 1957. Effects of selection for meiotic behaviour in autotetraploid rye. Hereditas, 43, 318-326.

Hossain, M.G., 1978. Effects of external environmental factors on chromosome pairing in autotetraploid rye. Cytologia, 43, 21-34.

Law, C.N., 1963. An effect of potassium on chiasma frequency and recombination. Genetica, 33, 313-329.

Lu, B.C., 1974. Genetic recombination in *Coprinus* IV. A kinetic study of the temperature effect on recombination frequency. Genetics, 78, 661-677.

Martin, A., 1978. Aneuploidy in *Vicia faba* L. The Journal of Heredity, 69, 421-423.

McCollum, G.D., 1958. Comparative studies of chromosome pairing in natural and induced tetraploid *Dactylis*. Chromosoma, 9, 570-582.

Morrison, J.W. and Rajhathy, T., 1960. Chromosome behaviour in autotetraploid cereals and grasses. Chromosoma, 11, 297-309.

Muntzing, A., 1936. The evolutionary significance of autopolyploidy. Hereditas, 21, 263-378.

Muntzing, A., 1951. Cytogenetic properties and practical value of tetraploid rye. Hereditas, 37, 17-84.

Pao, W.K. and Li, H.W., 1948. Desynapsis and other abnormalities induced by high temperature. Jour. Genet., 48, 297-310.

Poulsen, M.H. and Martin, A., 1977. A reproductive tetraploid *Vicia faba* L., Hereditas, 87, 123-126.

Riley, R., 1966. Genotype environmental interaction affecting chiasma frequency in *T. aestivum*. In: Chromosomes Today (C.D. Darlington and K.R. Lewis, eds., Vol. 1, pp 57-65. Edinburgh: Oliver and Boyd, 1966).

Riley, R., Chapman, V., Young, R.M. and Belfield, A.M., 1966. Control of meiotic chromosome pairing by the chromosomes of the homoeologous group 5 of *Triticum aestivum*. Nature, 212, 1475-1477.

Sjödin, J., 1971. Induced morphological variation in *Vicia faba* L., Hereditas, 67, 155-180.

Stebbins, G.L., 1947. Types of polyploids: their classifications and significance. Adv. Genetics, 1, 403-429.

Steffensen, D., 1957. Effects of various cation imbalance on the frequency of X-ray induced chromosomal aberrations in *Trasdentia*. Genetics, 42, 239-252.

Swaminathan, M.S. and Sulbha, K., 1959. Multivalent frequency and seed fertility in raw and evolved tetraploids of *Brassica campestris* var. Toria. Z. Vererbungsl,. 90, 385-392.

Sybenga, J., 1972. General Cytogenetics, North Holland, Amsterdam.

Tsuchiya, T., 1957. Fertility of autotetraploids and their hybrids in barley II. Meiosis and fertility in tetraploid hybrids. Seiken Ziho, 8, 27-32.

DISCUSSION

J. Picard *(France)*

Discussion is now open on the papers on Drs. Knudsen and Martin.

D.A. Bond *(UK)*

I have two questions for Dr. Knudsen. Did the value of 40 - 50% natural crossing in some open-flowered plants relate to a general figure for Denmark or was it a figure from the same field where the closed flower plants were grown and where 23% of crossing was obtained?

Would you agree, in view of the 20% crossing which was obtained on closed flower plants, we have now got to look for a combination of good auto-fertility in the closed flower character and that we should try to introduce a type of auto-fertility which operates early in the life of the flower?

J.C.N. Knudsen *(Denmark)*

In my figure of 40 - 50% I am referring to studies by Poulsen over a period of about two years. I do not have any figures for the field with normal flowers. But I must say that these were adequate for a very high crossing frequency.

I must agree with you that it is necessary to have a high degree of auto-fertility, and if it operated during early flowering it would help to reduce the crossing frequency.

M.H. Poulsen *(Denmark)*

I would like to add to that we did find a lot of variation among the families in terms of crossing. In fact we found a very low percentage of crossing in the material originating from crosses with the Sudanese Triple-White, which has an autofertility system that operates early in flower development.

M. Frauen *(FRG)*

I would like to make a comment on the closed flower mutant. We should not forget the advantage of partial outcrossing. The only advantage with the closed flower mutants and their self-fertilisation is that isolation chambers are not necessary, but hand pollination is required to make crosses. You can use the normal partial elogamy system for achieving many crosses with little effort. To improve the basic population of *Vicia faba* in our nurseries we use two methods, one with inbred lines and the other is to improve the total population by continual out-crossing.

G.F. Chapman *(UK)*

If you develop a closed flower variety which is subsequently distributed, have you locked yourself into a situation where, if you want to transfer a closed flower variety to something else, you prolong the amount of back crossing that must be done to stabilise it? Whilst it has certain attractions, one should not lose sight of the speed with which one can change plants. I wonder whether this is a cumbersome character.

D.A. Bond

I am not sure what Dr. Chapman means by ' a cumbersome character'. The attractiveness of closed flower to us is that it enables us to handle much larger amounts of material, and by so doing increases the probability of finding the recombinants that we want. At the moment we are still restricted in the number of lines we can handle.

J. Picard

If there are no more questions we will close the discussion.

RECOMMENDATIONS FOR FUTURE RESEARCH

G. Prendergast
Commission of the European Communities, DG VI,
200 rue de la Loi, Brussels, Belgium.

Ladies and gentlemen, I do not propose to keep you very long. The objectives of these meetings depend largely on what people want to get out of them. Firstly, it is for an exchange of information, especially between those who have contracts with the EEC, to give them an opportunity of exchanging information and, hopefully, we shall publish the results so that other interested parties within the Community can get that information.

Secondly, and just as important, it gives research workers in a particularly narrow field an opportunity of getting together. We have found, as a result of past meetings, that this leads to good cooperation and furthers actual programmes and research.

A third objective, which is not always part of the meeting, is to help the Commission to formulate future policy, possibly where a change of emphasis should be introduced. In this third area, conclusions from a meeting like this are very important. It is of great value if a few big questions arise from such a meeting. This is the point I would like to emphasise now: can you formulate important questions which would need research, or a future programme, to answer?

I have gained the impression from the last two days that the agronomists know the types of varieties, or cultivars, that they want, and want the breeders to produce them, but they have not described what is needed sufficiently well for the plant breeders. On the other hand, the breeders are of the opinion that they have the good varieties if only the agronomists treated them properly. Can you formulate questions in these two areas for future guidance?

There are a number of sister programmes, one being the Agrimed programme which may, in the next year or two, become a Big Brother to the protein programme. The main objectives are to help less favoured areas of the Community and naturally, being Agrimed, it relates to the Mediterranean area. As far as legume crops are concerned, two approaches are being considered. One is to grow legume crops for forage within the area, much of which suffers from water stress. Lupins would obviously be an important crop in such circumstances. But the crops which we are dealing with here today, *Vicia faba*, will also have a place. Irrigation too is being examined, so perhaps even what I call the 'Dantuma luxury-loving varieties' may have a place, but from what we heard yesterday maybe only a limited number can be involved in this. Therefore, this is a question for the plant breeders: can you think more along the lines of developing some varieties of *Vicia faba* more suited for the area that comes under the Agrimed programme. The second aspect of the work being considered concerns the warm, dry climate. It should be possible for them to produce seeds of most of the legume crops for use by farmers in northern Europe. Perhaps this is another area you may care to think about.

Perhaps you can formulate three or four questions for the plant breeders, the agronomists and their combined work over the next few years? We must think about a possible change of emphasis in the programme for the end of this year and 1982 and 1983. We must also think of the next 5 year programme, starting at the beginning of 1984, and we are already under some pressure to produce some broad outlines. This time next year, we must have a fairly definite plan. Thank you.

SUGGESTIONS FOR FUTURE POLICIES

D.A. Lawes (UK) suggested that everyone affirm their agreement with the overall protein production programmes and that *Vicia faba* had a role to play.

Dr. Prendergast thought it might be useful to put this in the report.

G.P. Chapman (UK), referring to the paper on flower and pod set from Durham, thought it valuable when people, who were doing a very practical plant breeding or agronomic job, were made to think about some fundamental aspect of physiology. He also pointed out a lack of commercial people at the meeting, feeling that they should be included more as their different priorities could be extremely stimulating, and that the group would benefit from being made more aware of commercial pressures.

D.A. Lawes (UK) remarked that the main problem with *Vicia faba* has been reliability of yield and that some re-modelling is necessary. In this context a knowledge of physiology, in particular, would be valuable. There should be close links between breeding and physiological studies. The two must be considered together.

R.B. Austin (UK) thought there was a good opportunity, at this early stage in the work on the field bean crop, for fruitful medium-term research, in terms of pay-off. Collaboration should be encouraged between breeders and physiologists to produce experimental genotypes differing in selected characteristics. The responses and consequences of those characters, and their genetic control, could then be explored and the information used to formulate better ideotypes and form the basis of breeding in the future. No single institute could do justice to all the various characters. Perhaps, in the context of EEC involvement, investigation of certain characters could be encouraged in particular research groups i.e. the determinate habit, or flowering synchrony, etc.

E. von Kittlitz (FRG) thought the variation found in the nurseries should be used more in developing varieties. Breeding methods should be explored to determine whether hybrids, synthetics or pure line varieties provided the best approach. It may be appropriate to include these investigations in contracts.

G. Dantuma (The Netherlands) thought studies were needed on water balance at the different stages of plant growth and development, in view of all that had been heard about drought stress. This could also be a short-term project.

J. Picard (France), who was also in the Chair, proposed that someone could be funded to take a sabbatical, visiting the various institutions in Europe working on *Vicia faba* and collecting information, to compile a report to the Commission on suggestions for the direction of *Vicia faba* research relevant to the interests of the Commission. He thought 15 - 20 minutes of discussion was insufficient to answer all Dr. Prendergast's questions adequately. There was general assent and Dr. Picard suggested that the proposal be sent to DG6 who could approach all the institutes asking for someone who could give the time.

Dr. Prendergast thought that the proposal could possibly be made towards the end of this year.

M. Frauen (FRG) thought that plant breeding programmes on *Vicia faba* should be speeded up. Breeding is, by its nature, a long-term process, but it could be speeded up by using two generations a year. Perhaps the EEC could provide for a second generation to be produced in the southern hemisphere for breeders involved in the crop. (There was general assent).

R. Thompson (UK) agreed with the proposal put by Dr. Dantuma but felt it should include reduction of stress in relation to other factors i.e. edaphic factors such as pests and diseases and nutrients. This could be done in such a way as to provide information on genotype performance in relation to drought tolerance and also to the local aerial environment - radiation, daylength and temperature.

D.A. Lawes (UK) suggested that a special clause be included that, in this meeting, the consensus has been that the availability of soil moisture, or moisture in general, is so important in this crop that it is a special area of work.

A.F. van der Wal (The Netherlands) suggested that the different topics examined at this meeting should be integrated. For example, the work of Gates on flower morphology and anatomy relates to hormone balance and the physical aspects of plant growth, especially water potentials in the soil and in the plant, and also to temperature and radiation. He wanted to see the development of screening methods which could cope with large numbers of plants, and which were economical and also relevant. Many screening tests were really only marginally relevant to the breeders' needs.

M. Frauen (FRG) thought that seed quality (in relation to germination, vigour and performance) in legumes was a special problem. Ideal locations should be sought for seed multiplication He thought that the drier areas were more suitable for seed production. Perhaps an examination of this could be linked with the joint field bean tests which were testing varieties in a wide range of climates.

M.H. Poulsen (Denmark) spoke from the point of view of someone in the private sector. He did not think that breeders in Denmark would like to see seed multiplication done anywhere but in Denmark; the Dutch breeders would like seed multiplication done in Holland - for 'obvious reasons', he said.

Jennifer Yarwood (UK) suggested that much value could be derived from a list detailing the range in each of the many characters found in *Vicia faba*, not only for commercial varieties but also for genotypes perhaps discarded by breeders. Speaking as a physiologist, she said it was difficult to locate particular characters when they were required.

G.P. Chapman (UK) said that, in fact, he had compiled such a list. It had been compiled on the basis of whether it was a whole plant character, a leaf character, stem, seed, disease resistance, chromosome number, etc. It had been compiled 2 years ago. He thought it would be helpful if it could be circulated through members of this meeting, indicating where there had been changes i.e. where there had been one mutant before, there were now six. It was time that his list was updated, and although it was of limited usefulness it was better than nothing.

D.A. Lawes (UK) wondered if the list could be included in the 1980 report of the meeting.

G.P. Chapman (UK) did not feel that it was sufficiently up-to-date to be included. He added that if alterations were to be made, a closing date should be decided upon.

J. Picard (France) suggested that it might be published in Fabis.

G.C. Hawtin (Syria) said that, to be published in Fabis, it should be with them by the end of the year and could then go out in Fabis No. 3. but it would need to reach Icarda by the end of December. Rather than have the list pass from hand to hand, he volunteered to circulate copies to everyone at the meeting, plus key people who may be able to add their contributions. This list could then be returned to Dr. Chapman.

G.P. Chapman (UK) agreed to be editor of the list.

B. Snoad (UK) said that, as far as peas were concerned, a similar system had been operating for some years and nothing but good had come from it. He said he could recommend it wholeheartedly on the basis of practical experience.

LIST OF PARTICIPANTS

R.B. AUSTIN
Plant Breeding Institute,
Maris Lane,
Trumpington,
Cambridge CB2 2LQ,
UK.

W.H. BAIER
Fa. Pflanzenzucht Franck,
Oberlimpurg,
7170 Schwabisch Hall,
FRG

D.A. BOND
Plant Breeding Institute,
Maris Lane,
Trumpington,
Cambridge CB2 2LQ,
UK.

P. BERTHELEM
Station d'Amélioration des Plantes,
BP 29,
35650 le Rheu,
France.

G.P. CHAPMAN
Wye College,
Nr. Ashford,
Kent,
UK.

J.J. CUBERO
Dept. de Genetica,
Apartado 246,
Cordoba,
Spain.

G. DANTUMA
Centre for Agrobiological Research,
Bornsesteeg 65,
Postbus 14,
6700 AA Wageningen,
The Netherlands.

H.M. DEKHUIJZEN
Centre for Agrobiological Research,
Bornsesteeg 65,
Postbus 14,
6700 AA Wageningen,
The Netherlands.

G. DUC
Station d'Amélioration des Plantes,
INRA,
BV 1540,
21034 Dijon Cedex,
France.

M. FRAUEN
Inst. für Pflanzenbau und Pflanzenzüchtung,
von Sieboldstrasse 8,
34 Göttingen,
FRG

P. GATES	Botany Department, University of Durham, Science Site, South Road, Durham, UK.
J. le GUEN	Station d'Amélioration des Plantes, BP 29, 35650 le Rheu, France.
G.C. HAWTIN	Icarda, Box 5466, Aleppo, Syria.
C.L. HEDLEY	John Innes Institute, Colney Lane, Norwich NR4 7UH, UK.
R.J. HERINGA	Foundation for Agricultural Plant Breeding, Droevendaalsesteeg 1, Postbus 117, 6700 AC Wageningen, The Netherlands.
E.R. KELLER	Institut für Pflanzenbau, ETH Zentrum, Universitätstrasse 2, 8092 Zürich, Switzerland.
E. von KITTLITZ	Landessaatzuchtanstalt, Universität Hohenheim, Postfach 106, 7000 Stuttgart 70, FRG
J.C.N. KNUDSEN	K.V.L. Landbrugets Plantekultur, Agrovej 10, 2630 Tastrup, Denmark.
D.A. LAWES	Welsh Plant Breeding Station, Plas Gogerddan, Near Aberystwyth SY 23-3 EB, UK.
G. LOCKWOOD	Plant Breeding Institute, Maris Lane, Trumpington, Cambridge CB2 2LQ, UK.

A. MARTIN	Dept. de Genetica, Apartado 246, Cordoba, Spain.
D. MELIN	Laboratoire Botanique Institut Sc. naturelles, 25042 Besancon-Cedex, France.
G. MOSCA	Istituto di Agronomia, Via Gradenigo 6, 35100 Padua, Italy.
K. NAGL	Bundesanstalt für Pflanzenbau und Samenprüfung, Alliertenstrasse 1, 1201 Vienna, Austria.
ABDULLAH NASSIB	Grain Legume Research Station, Field Crops Research Institute, Agricultural Research Centre, Giza, Egypt.
G. NILSSON	Svälof AB, S 26800 Svälof, Sweden.
J. PICARD	Station d'Amélioration des Plantes, INRA, BV 1540, 21034 Dijon, France.
M. POISSON	Station d'Amélioration des Plantes Fourragères, 86600 Lusignan, France.
G. POMMER	Bayerische Landesanstalt für Bodenkultur und Pflanzenbau, Vöttingerstrasse 38, D 8050 Freising, FRG
M.H. POULSEN	Dansk Planteforaedling A.S., Boelshøj, DK 4660 Store-Heddinge, Denmark.
J. PUECH	Station d'Agronomie, INRA, BP12, 31320 Castanet Tolosan, France.

G.G. ROWLAND

Crop Science Department,
University of Saskatchewan,
c/o Plant Breeding Institute,
Maris Lane,
Trumpington,
Cambridge CB2 2LQ
UK.

GABRIELE RUSITZKA

Institut für Nutzpflanzenforschung,
T.U. Berlin,
Albrecht-Thaer-Weg 5,
1000 Berlin 33
GDR

M.C. SAXENA

Icarda,
PO Box 5466,
Allepo,
Syria.

W.P. SCHRÖDER

Institut für Nutzpflanzenforschung,
T.U. Berlin,
Albrecht-Thaer-Weg 5,
1000 Berlin 33,
GDR

J. SJÖDIN

Svälof AB,
S26800 Svälof,
Sweden.

B. SNOAD

John Innes Institute,
Colney Lane,
Norwich NR4 7UH,
UK.

J.H.J. SPIERTZ

Centre for Agrobiological Research,
Bornsesteeg 65,
Postbus 14,
6700 AA Wageningen,
The Netherlands.

R. THOMPSON

Scottish Crop Research Institute,
Invergowrie,
Dundee DD2 5DA,
UK.

A. PH. de VRIES

Instituut voor Plantenveredeling,
Lawickse Allee 166,
Wageningen,
The Netherlands.

A.F. van der WAL

Foundation for Agricultural Plant Breeding,
Droevendaalsesteeg 1,
Postbus 117,
6700 AC Wageningen,
The Netherlands.

G. WALTON — Department of Agriculture,
South Perth 6151,
West Australia.

F. de WOLFF — D.J. van der Have,
Plant Breeding Station,
PO Box 1,
4410 AA Rilland
The Netherlands.

JENNIFER YARWOOD — Dept. of Botany,
University of Durham,
Science Laboratories,
South Road,
Durham DH1 3LE,
UK.

U. ZILIOTTO — Istituto di Agronomia,
Via Gradenigo 6,
35100 Padua,
Italy.

RECORDING PERSONNEL

GILLIAN COOKES
S.E.W. HALLAM

Janssen Services
14 The Quay,
Lower Thames Street,
London EC3R 6BU
UK.

Manuscript was prepared by:

Janssen Services, 14 The Quay, Lower Thames Street, London EC3R 6BU, UK